Development, Poverty, and Power in Pakistan

Rural development remains a major challenge for governments of developing countries such as Pakistan. While a broad range of state and donor interventions impact the lives of poor farmers – who provide a significant proportion of the labour force – comprehensive consideration of these combined interactions remains inadequate. Focussing on Pakistan, this book discusses the political economy of agrarian poverty and underdevelopment in the region.

The book provides an in-depth exploration of the combined impact of state and donor interventions, as well as that of resistance attempts, to alter the status quo within Pakistan. It questions the relevance of state institutions and policies contending with the problems of farmers in Pakistan, and how donor-led policies and programmes also influence their lives. It draws on findings that have emerged from interviews of over 200 respondents including government officials, donor agency representatives and different categories of poor farmers, during eleven months of fieldwork in the provinces of Sindh and Punjab. This research reveals some divergences between state and donor policies, but it finds more prominent convergences, which in turn enable the landed rural elite to benefit from market-based and capital-intensive processes of agricultural growth, without offering substantial opportunities for poor farmers.

Reflecting the need to become less insular when discussing solutions to rural development, this book will be of interest to students and scholars of South Asian Politics and Development Studies.

Syed Mohammad Ali is a post-doctoral fellow at McGill University, Canada. He holds a PhD in development anthropology and has worked in the development sector in Pakistan for fifteen years. Besides academic writing, he has contributed a weekly op-ed column in Pakistani newspapers for the past ten years.

Routledge Contemporary South Asia Series

Development, Poverty, and Power in Pakistan

The impact of state and donor interventions on farmers

Syed Mohammad Ali

Routledge
Taylor & Francis Group

LONDON AND NEW YORK

First published 2015
by Routledge
2 Park Square, Milton Park, Abingdon, Oxon OX14 4RN

and by Routledge
711 Third Avenue, New York, NY 10017

First issued in paperback 2017

Routledge is an imprint of the Taylor & Francis Group, an informa business

British Library Cataloguing in Publication Data
A catalogue record for this book is available from the British Library

Library of Congress Cataloging in Publication Data
Ali, Syed Mohammad, author.
  Development, poverty and power in Pakistan : the impact of state and
donor interventions on farmers / Syed Mohammad Ali.
     pages ; cm. -- (Routledge contemporary South Asia series ; 94)
  Includes bibliographical references.
  1. Farmers--Pakistan--Economic conditions. 2. Rural poor--Pakistan.
3. Rural development--Pakistan. 4. Agriculture and state--Pakistan. I. Title.
II. Series: Routledge contemporary South Asia series ; 94.
  HD8039.F32P182 2015
  338.1095491--dc23
                2014029506

ISBN 13: 978-1-138-49228-8 (pbk)
ISBN 13: 978-1-138-80453-1 (hbk)

Typeset in Times New Roman
by Taylor & Francis Books

For my parents, and my son, Zameer

Contents

Acknowledgements

It is a pleasure to acknowledge several people who made this research possible. I would like to thank my Ph.D. supervisor, Dr Nadeem Malik, for his work on my Ph.D. thesis, the findings of which are drawn upon in the writing of this book. I am also grateful to the Open Society Institute in Budapest for providing me a two-year fellowship to conduct research on the Poverty Reduction Strategy Paper for Pakistan, and its impact on poor farming communities. In Pakistan, I would like to acknowledge several development organizations (including the Thardeep Rural Support Programme, the National Rural Support Programme, the International Labour Organization, ActionAid) and academics (especially Dr Imran Ali and Dr Ahsan Rana) whose insights and support helped me access several of the field sites where this research was conducted, and to identify issues and respondents of relevance to my research. I am also grateful to my wife, Bariza Umar, for her intellectual input and support during the most arduous stages of writing. Most of all, I am indebted to my numerous informants for providing me their views, without which this book could not have been written.

Note on conversions

Pakistani rupee	the Pakistani Rupee average conversion rate to the United States dollar (USD) was Rs 100.98 to USD1 during the first six months of 2014, and its conversion to the Great Britain pound (GBP), over the same period, was Rs 168.44 to GBP1
acre	1 acre is equivalent to 0.405 hectares
kanal	there are 20 kanals in one acre
marla	there are 20 marlas in one kanal
maund	there are 40 kilograms in one maund

Abbreviations

ADB	Asian Development Bank
ADBP	Agricultural Development Bank of Pakistan
AHRC	Asian Human Rights Commission
AIT	Agricultural Income Tax
CFO 2001	Corporate Farming Ordinance 2001
DFID	Department for International Development
FBR	Federal Board of Revenue
FGDs	focus group discussions
FOs	Farmer Organizations
GDP	gross domestic product
GMO	genetically modified organism
GoP	Government of Pakistan
HDI	Human Development Index
HDR	Human Development Report
IMF	International Monetary Fund
KPK	Khyber Pakhtunkhwa Province
LRMIS	Land Record Management System
MHHDC	Mehbub-ul-Haq Human Development Centre
MNCs	Multinational Corporations
MQM	Muttahida Qaumi Movement
NFML	National Fertilizer Marketing Limited
NGO	non-government organization
NIPS	National Institute of Population Studies
NWFP	North West Frontier Province [renamed Khyber Pakhtunkhwa Province]
OECD	Organisation for Economic Co-operation and Development
PBR Bill 2012	Plant Breeders' Rights Bill 2012
PFF	Pakistan Fisher-folk Forum
PIDA	Provincial Irrigation and Development Authorities
PIDE	Pakistan Institute of Development Economics
PML (N)	Pakistan Muslim League (Nawaz)
PPAF	Pakistan Poverty Alleviation Fund
PPP	Pakistan Peoples' Party

QUANGOs	quasi-autonomous non-government organizations
RSPs	rural support networks
SDPI	Sustainable Development Policy Institute
TCP	Trading Corporation of Pakistan
TRIPs	Trade Related Aspects of Intellectual Property Rights
UAE	United Arab Emirates
USAID	United States Agency for International Development
WTO	World Trade Organization

1 Introduction

According to recent estimates, about 75 per cent of the world's poorest people live in rural areas and are involved in agricultural activities, which comprise their main source of employment and income (IFAD 2013). Given this basic fact, the need to address the plight of poor farmers remains a major challenge for governments of developing countries, as well as the international donor agencies that support them.

Effectively addressing the concerns of poor farmers, however, requires wide-ranging efforts, which transcend the confines of agricultural development policies, or even state- or market-led agrarian reforms. This is because a much wider range of state institutions and policies, as well as donor-endorsed development policies and programmes, have a direct impact on the lives of poor farmers. Broad legislative processes and institutions, like the judiciary and legislature, the military, or even the police, play a very important role in shaping circumstances faced by poor farmers in countries like Pakistan. Similarly, overarching donor-supported policy frameworks such as the Poverty Reduction Strategy Papers (PRSP),[1] which emphasize the need for increased privatization, liberalization, and the adoption of market-based mechanisms to achieve economic growth in developing countries, also have significant implications for poor farmers.

Moreover, in the era of globalization, contending with problems-facing marginalized groups like poor farmers requires cognizance of production relations at not only the national, but at the international level as well (Lodhi and Cristóbal 2010). Yet, scholars like Byres and Bernstein (2001) caution against tendencies of looking for universal accounts explaining the totality of social and political implications related to contemporary capitalism and its relationship to the countryside. This is understandable, given that larger configurations of agricultural production on a global scale are translated into production arrangements within particular rural economies, which have context-specific implications that cannot be generalized. Grappling with such issues requires cognition of the individual nature of specific state formation processes, as well as how a wide range of state institutions are interacting with location-specific rural communities based on their own, and/or donor-prescribed notions, of what it means to pursue goals such as agricultural growth or rural poverty alleviation. Yet avoiding generalizations is often difficult in practice, often leading to the adoption of 'one-size-fits-all' policy prescriptions.

Much of the current debate about how to address problems confronted by poor farmers also tends to become rather insular. For instance, World Bank economists, Dorosh and Salam (2007), in looking for justifications to dissuade developing countries from providing agricultural subsidies, point out that many of the benefits of subsidization are being availed by large landowners rather than poor farmers, due to corruption within state agencies. However, this analysis does not draw attention to what donor agencies can, or should, do to ensure that governments in developing countries effectively target subsidies to poor farmers. Similarly Easterly (2001), a prominent World Bank economist, while discussing reasons for the failure of financial resources and technical advice provided by the World Bank in improving the lives of the poor, places the blame on the governments of developing countries and on their political elite, without adequate consideration of the role of donor agencies in implicitly aggravating these problems. Scholars such as Zhaung, Dios, and Lagman-Martin (2010), Kaufmann *et al.* (2009) and Brooks *et al.* (2010), have also failed to sufficiently address the combined impact of state and donor policies and programmes on poor farmers in developing countries. For example, they have prescribed the need for rule of law, political stability, bureaucratic capability, and control over corruption, which puts a disproportional onus onto developing countries' governments, for the achievement of pro-poor growth.

On the other hand, criticism concerning the World Bank's ability to deliver pro-poor agricultural development often zooms in on its ideological predisposition towards the market mechanism, without sufficient exploration of how particular vested interests within different states choose to adopt, manipulate, or reject different policies and programmatic interventions. Stein (2010), for instance, challenges the logic underlying the World Bank's liberalization policies in sub-Saharan Africa, and largely blames the imposition of market mechanisms for remaining unable to achieve income equality and reduce poverty. While he recognizes that part of this failure is embedded in the distorted access to agricultural inputs and lack of access to markets, there is insufficient analysis of how vested interests within the state are creating such hurdles, or how and why different state institutions and policies interact with particular World Bank programmatic interventions to reinforce rural inequalities.

There is ample evidence of lop-sided analyses within developing countries themselves. For instance, Chaudhry *et al.* (2006), while trying to explain reasons for the phenomenon of rural poverty, point to how agricultural development efforts preserve the interests of the landed elite, but then go on to blame inept state policies with regard to agricultural taxation, price support and subsidies, and an excessive regulation of the agricultural marketing system. Nowhere in this analysis is the role of donor agencies in exacerbating rural inequalities considered. In another influential study, Anwar *et al.* (2004) correlate the incidence of rural poverty to lack of landownership, to make a convincing argument for land redistribution in Pakistan. However, the adverse impact of donor-supported policies, such as boosting productivity through capital-intensive farming or corporate farming, which provides impetus for large-scale farming, is not taken into account. Conversely, Khan (1999) and Gera (2004) have argued that market-led reforms, such as the liberalization of agriculture, fail to address challenges confronting poor farmers, including women, but they have not drawn much attention to which vested interests within the state diverge from, or reinforce, this process.

Nonetheless, in countries like Pakistan, an increasing convergence is occurring between local (rural elite, politicians, and military), and international (trade, corporate, and donor) interests concerning the means by which to pursue agricultural growth (Gazdar 2005; Siddiqa 2007). Still, many recent studies focusing on different aspects of agricultural development and rural poverty related challenges do not adequately deal with the combined impacts of current state and donor policies on poor farmers. Recent research undertaken in rural Punjab (Cheema *et al.* 2009), demonstrates continuing monopolization of state resources by the landed rural elite, who had been granted unequal political and economic power by the colonial state prior to independence. However, this assessment does not explore if, and how, current donor policies or programmes are helping the historic landed elite in rural Punjab maintain its hold on power.

Even when there is recognition of parallel state and donor policies adversely affecting poor farmers, the specific interconnections between state and donor policies are not sufficiently explored. Toor (2010), for instance, criticizes the Pakistani state's reluctance to redistribute agricultural land to poor farmers, which, alongside the World Bank's encouragement of capital-intensive export-oriented production, is considered to be undermining household food security in rural areas. Toor does not, however, explore which particular vested interests are benefiting from the existing donor endorsement of promoting export-oriented agricultural policies, or what enabling role is played by different state institutions in this process. Similarly, the fact that market mechanisms can themselves become asymmetrical in rural areas is corroborated by empirical evidence provided in the National Human Development Report in 2003 (Hussain *et al.* 2003), whereby poorer farmers have to pay more for inputs and get less for their outputs in comparison to large landowners. Yet, how the combination of local power dynamics interacts with different institutions of the state, and how specific donor policies aim to address or exacerbate these market distortions, is not well understood.

Much of the existing emphasis on state or donor policies has also failed to adequately recognize the potential role of poor farmers in addressing their own problems. The significance of peasant movements is usually highlighted based on the assumption that state and market mechanisms can be held captive to vested interests, which prevents them from carrying out significant pro-poor reforms in rural areas (Borras *et al.* 2008). Although they have grown in numbers and strength, peasant movements still fall prey to internal disputes and find it difficult to link with broader reformist attempts for achieving transformative change (Borras, Edelman, and Kay 2008; Lindemann 2010). What context-specific impediments continue to undermine peasant movements from effectively achieving the goal of empowering poor farmers is an important issue, which needs further consideration. In this context, understanding emerging ruptures within ongoing resistance movements can promote lessons relevant for the ongoing theoretical and policy discussions concerning peasant movements.

In the case of Pakistan, the military is a major agricultural landowner. Aiming to maximize profits, the military tried to evict sharecropping tenants from agricultural lands controlled by it in order to lease out these lands, which in turn sparked a resistance movement by landless farmers who had been working on these military lands for generations. Some of the emergent literature on this resistance movement, including studies by Akhtar (2006) and Zaidi (2012a), acknowledge the role of colonial land grant schemes in enabling the military to gain control over agricultural lands, but thereafter emphasis is primarily placed on the prevailing military domination within the Pakistani state, and its coercive tactics to exploit poor farmers. Such analysis does not draw sufficient attention to how military farms have also been influenced by donor-endorsed policies, which have led to the increasing commercialization of tenancy contracts. There is also the need for critical assessment of civil society organization attempts to support this resistance movement in challenging prevalent institutional or policy discourse hegemonies, which exploit poor farmers.

Overall, significant gaps remain in terms of understanding how individual state institutions and the policies implemented by them engage with corresponding donor policies and programmes, and how and why they converge or diverge. In turn, what consequences this combined interaction has on poor farmers in developing countries, and to what extent resistance by poor farmers is capable of altering their existing circumstances, also needs further exploration. In the effort to identify linkages between these varied issues, this book draws attention to a wide range of state- and donor-led efforts, which have salient implications on the lives of poor farmers in a developing country like Pakistan. An attempt is made to highlight how state institutions choose to resist certain aspects of donor advice such as curbing public spending in the agricultural sector, primarily to preserve the vested interests of the landed rural elite. Besides emphasizing such divergences, a much broader convergence of interests between state and donor policies is identified, which enables the landed elite to benefit from capital intensive and growth-led agricultural production strategies endorsed by development frameworks promoted by major donor agencies. Resistance by landless farmers to the increasing commercialization of land markets is also explored, especially to draw lessons from the failure of resistance by farmers to challenge the hold of the military over agricultural land in rural Pakistan. This simultaneous analysis, which compares and contrasts a variety of relevant state and donor interventions of relevance to poor farmers, and the problems faced by resistance attempts to effectively alter the prevailing status quo in rural Pakistan, thus serves to address the above-identified gaps concerning prevalent challenges that confront poor farmers.

Scope and aim of the book

An important element for understanding problems facing poor farmers within the specific context of contemporary Pakistan requires the adoption of a comprehensive perspective. Such a perspective needs to simultaneously consider how different state institutions and policies interact with ongoing donor policies and programmes, so as to preserve the vested interests of the landed rural elite, despite attempts by poor farmers to resist the prevailing status quo. This book focuses on these issues, by posing the following three research questions:

1 How are relevant state institutions and policies contending with the problem of poor farmers, particularly in Sindh and Punjab?
2 How are major donor-led policies and programmes influencing the lives of poor farmers in Sindh and Punjab?
3 Why do recent instances of resistance by poor farmers lack the capacity to overcome the underlying causes of their marginalization?

These above questions are answered by focusing on a range of existing state policies and institutions, and donor-funded programmes, which exert direct influence on poor farmers in the country, as well as by analysing the interconnections between these varied attempts.

Besides focusing on state-led reforms, which claim to specifically focus on improving the lives of poor farmers, the roles of other state institutions such as the judiciary, the legislature and different branches of the executive are considered as they pertain to directly impacting the situation of poor farmers. While this book mentions some of the prominent bilateral development agencies working in Pakistan, it primarily focuses on World Bank policies and programmatic interventions, which exert overarching influence on formulating development policies in Pakistan. Specific attention is drawn to World Bank-influenced policies like the PRSP for Pakistan, and to specific programmatic interventions supported by this strategic framework, in the effort to achieve agricultural development and alleviate rural poverty. In addition to demonstrating how and why state and donor attempts collectively remain unable to effectively help poorer farmers, a resistance movement launched by landless farmers working on military controlled farms in the Punjab is also examined to understand why it remains unable to challenge the prevailing landownership patterns in the country, which are a major cause of the lingering rural deprivation.

Methodological approach

Besides collation and analysis of secondary data, including a broad range of policy and programmatic documents, field data were gathered from different locations in two provinces of Pakistan, Sindh and Punjab. These primary data were obtained from interviews with over 230 respondents during 11 months of research conducted in 2011. However, this book also takes note of subsequently relevant developments within the country as they kept unfolding until the general elections in May 2013.

After providing a brief rationale for selecting particular field sites within Sindh and Punjab, and a description of research tools, further information is provided concerning the research respondents themselves. Thereafter, some of the challenges faced in conducting this research, and strategies adopted to overcome them will also be mentioned.

a. Rationale for selection of field sites

One reason for the selection of field sites within the provinces of Sindh and Punjab was because 58 per cent of the total (177 million) population of the country lives in Punjab alone, followed by Sindh, which is home to 23 per cent of the country's population (Federal Bureau of Statistics 2011; NIPS 2008). The high degree of landlessness within both these provinces was another factor. The distribution of land holdings at the provincial level indicates that about 85 per cent of households own no land in Sindh, which is the highest proportion of landlessness among all the provinces in the country. Landlessness in Punjab is also very high, with 74 per cent of the population having no access to land (Anwar *et al.* 2004). Besides the problem of landlessness, there are many large landowners in both the provinces. Merely 0.01 per cent of households own more than 2 hectares (4.9 acres) of land in Sindh, followed by 0.2 per cent in the Punjab, suggesting a highly skewed landownership pattern (Anwar *et al.* 2004: 865). Such skewed landholding patterns make both Punjab and Sindh very relevant places to examine the implications ongoing state and donor attempts to address problems facing poor farmers.

Poverty in rural areas of both Sindh and Punjab is inextricably linked to land tenure arrangements, which have been actively shaped through colonial and post-colonial regimes, as well as donor interventions subsequent to the emphasis on 'Green Revolution' technologies in the 1970s, and the ongoing emphasis on market orientation of the agricultural sector. Despite considerable changes over these past few decades in the means of agricultural production, such as the use of high yielding crop varieties, increased prices of inputs, and growing mechanization, landlord–tenant relations are described as continuing to exhibit feudal dynamics within both these provinces, especially in districts where landholding patterns remain very uneven (Cheema *et al.* 2008, 2009). Focusing on these two provinces, therefore, enables exploring how the post-colonial state institutions/policies and donor policies and programmes on poor farmers.

More specific site selection within these two provinces was undertaken to ensure that relevant data could be obtained concerning the three main research questions posed in this book. Therefore, research done in Lahore (the capital city of the province of Punjab) was meant to gather information from senior government and donor agency representatives, including civil society organizations working with them, concerning relevant institutional mechanisms, policies, and programmes of most relevance to the lives of poor farmers. Fieldwork was conducted in the Gujrat district in the province of Punjab, and in the Umerkot and Badin districts in the province of Sindh to assess the impact of the relevant institutions, policies, and programmes on poor farmers. In the Umerkot and Badin districts in Sindh, the implementation of an ongoing government-led land distribution scheme, intended to benefit poor landless farmers, was assessed. Research in another district in southern Punjab[2] particularly aimed to understand the impact of a prominent corporate farm on poor farming communities in its vicinity. Moreover, research in the Okara district in Punjab provided insights concerning a prominent resistance movement by landless farmers.

b. Research tools

Research presented in this book is qualitative in nature and it does not depend on a statistical sampling of data. Instead, qualitative interviewing is used since it provides a flexible and powerful tool to obtain data in the social sciences (Rabionet 2011). Such methods are often preferred since they provide an opportunity to obtain both depth of information, and to resolve seemingly conflicting information, because the researcher has the opportunity to interact with respondents (Harrell and Bradley 2009). This research particularly relies on semi-structured interviews and focus group discussions.

Using semi-structured interviews enables specific themes to be predetermined, but interview formats can remain flexible (Kvale 1996). In contrast to completely unstructured interviews, using the semi-structured format allows exploring topics and themes closely related to the research questions under consideration, while at the same time enabling respondents to provide additional information of particular relevance to them (Rabionet 2011). The need for in-depth exploration of major issues concerning the research questions requires a free-flowing conversation around the theme, rather than asking a set of preselected questions. Semi-structured interviews also help determine emphasis by asking individual respondents how strongly they feel about certain issues, and to even ask respondents to prioritize issues or policies (Harrell and Bradley 2009). Semi-structured interviews were thus used to ascertain the priorities of different types of stakeholders, including senior government and donor agency officials and civil society representatives. Semi-structured interviews also proved useful for obtaining data from land revenue, agriculture, irrigation, and police department officials at the district level, as well as from several other relevant stakeholders such as seed and fertilizer dealers and commission agents. The impacts of state institutions and policies, and donor policies and programmes, on poor farmers were explored using semi-structured interviews at the village level as well.

Despite being able to obtain significant data through semi-structured interviews, this research method is constrained by its inability to provide information or opinions from primarily an individual perspective. To obtain a broader perspective concerning issues of relevance to my research, focus group discussions (FGDs) were conducted, which are particularly useful for obtaining diversified views of respondents on a given thematic issue. FGDs are not as relevant to determining emphasis, since group members do not often share the same emphases, and group dynamics often make it difficult to articulate a response that is accurately representative of the entire group (Harrell and Bradley 2009). Yet they remain a particularly appropriate research tool for unveiling and understanding the experiences of vulnerable populations, where the power of group interaction, enabled by focus groups, facilitates the disclosure of opinions, attitudes, or behaviours that may normally be kept private (Kulavuz-Onal 2011). FGDs were, however, held with different categories of stakeholders individually to respect the need for privacy, and to help create a secure environment for the more vulnerable category of stakeholders (including sharecroppers, women, and agricultural labourers) to share their views. FGDs with different categories of poor farmers helped obtain information concerning issues of particular significance for them as a specific group, such as their views on inheritance rights, remuneration, and conditions in which they work, and to assess the effectiveness of state or donor policies in addressing their needs. Some of these FGDs were in mixed gender groups, especially in villages in Sindh, where gender segregation was not being very strictly observed within the communities themselves, or where members of the group were related to each other, and therefore felt comfortable expressing their views in a group.

Participant observations were used during the fieldwork, which are typically employed to undertake qualitative sociological and anthropological research (Gillespie and Michelson 2011). The use of participant observations proved very useful for this research, since it provided information based not only on what different respondents said, but also how they behaved and interacted with other in a particular social environment. Using the participant observation technique required more than just using fieldwork based recollections as anecdotes to add richness to my research. Instead, use of such an ethnographic method required rigorous and labour-intensive work. After each field trip, I had to systematically write down every event or occurrence that had taken place during the course of the field trip. This extensive record has in turn enabled me to find certain patterns and make inferences, which are presented in subsequent sections of my findings. Participant observations were conducted in villages and the offices of relevant local government officials, and exclusively involved respondents who had already consented to be part of this research. Such participant observations have helped create an awareness of the social environment of poor farmers, and to obtain insights concerning the interaction and attitudes of government officials towards poor farmers.

The real names of respondents who consented to participate in this research have not been revealed. The anonymity of research respondents is maintained in view of ethical concerns involved in conducting social research, which aims to ensure that research findings do not adversely affect the feelings or lives of those studied (Giddens 1991). In view of these considerations, respondents are either not named, or their names have been altered. However, in order to provide enough specificity to help contextualize data sources, research findings have been attributed to respondents by reference to their designations (a senior official in the Irrigation department, for instance), without revealing details that can identify individuals. Similarly, the geographic locations where particular research findings were obtained have been indicated. However, in the spirit of preserving a required degree of anonymity, the name and location of a specific corporate farm, whose working have been discussed in detail in later chapters, has not been revealed.

c. Data sources

The primary data discussed in this book are based on feedback from a wide range of stakeholders including policy makers, programme implementers, civil society organizations, independent analysts, and intended beneficiaries of state and donor development interventions, such as landless women and several categories of poor farmers.[3]

Interviews were conducted with government officials and other stakeholders (commission agents, activists, etc.) at the district level, and then with farmers within selected village communities. In each village, different categories of farmers were interviewed (landowners with varying sizes of land holdings, sharecroppers owning some land in addition to sharecropping, sharecroppers working on different sizes of land, landless farmers, and women engaged in farming activities).[4] Interviewing these different categories of farmers was necessary to indicate a range of variations and consistencies within their perspectives and circumstances.

The primary aim of choosing respondents was to obtain feedback from all the categories of respondents mentioned above, rather than achieve a statistical balance across different field sites or even between the different categories of respondents. Therefore, the preference was for purposive sampling, whereby the choice of respondents was guided by their informational value.[5]

In total, 234 (173 male and 61 female) respondents were interviewed for this research. Over 150 of these respondents were engaged in farming themselves, while the remaining were government officials, donor agency representatives, or civil society representatives.

While some of the semi-structured interviews conducted during the fieldwork lasted for only approximately half an hour, more of them carried on for at least an hour, while those with key informants lasted longer. Around a dozen of the key informants, including a local land revenue official in Gujrat, and activists and several villagers in all the districts where research was conducted, were willing to participate in multiple interviews during follow-up visits to their localities.

Access to respondents was obtained using varied strategies. Having worked in the development sector in Pakistan for 15 years, I had cultivated strong contacts with several local and international NGOs, donor agencies, and government officials working on rural poverty issues across the country. My links with these different stakeholders proved extremely helpful in obtaining information, relevant documents, and scheduling appointments with other relevant respondents in the government, donor, and NGO community. NGOs working on capacity building or empowerment projects also helped introduce me to village communities in both Sindh and Punjab.

Besides research done in the two provinces, secondary data was obtained from books, scholarly articles, NGO reports, government reports, and other documents. Recent legislation, concerning corporate farming, patent rights, and a bill aiming to protect women from being denied landownership, has also been reviewed, alongside public interest litigation attempts to instigate land reforms, and election reforms to curb the influence of the landed rural elite in politics.

d. Challenges and strategies used to overcome constraints

Given the need to obtain information from this diverse range of respondents, I did face some challenges in conducting my fieldwork and had to adopt different strategies to overcome the emergent constraints.

Interviews with some categories of respondents were fairly easy to conduct, with interviewees giving me as much time as I needed. Getting appointments with some of the senior government officials, however, took several scheduling requests, and the interviews were over within a matter of a few minutes due to the busy schedules of these respondents. With interviewing these officials, I made it a point to commence the semi-structured interviews by posing open-ended queries of direct relevance to their own area of responsibility, and then, if time allowed, I would move on to discussing broader issues.

Some of these respondents were well educated and spoke multiple languages with ease, while others were restricted to communicating in their regional language, without recourse to Urdu, the national language of Pakistan, nor English, a language entirely foreign to them. As a 'local researcher', born in the city of Lahore and a native speaker of Punjabi and Urdu, I did not have trouble communicating with unilingual locals in Punjab. I was thus able to approach village communities directly, create personal links, and start to carry out the fieldwork. In Sindh, I was also helped by field officers affiliated with a local NGO (Tardeep Rural Development Programme) to not only reach out to remote villages, but also communicate with some of the villagers who only spoke their provincial language (Sindhi). There were, however, several village-based respondents who also spoke the national language (Urdu), with whom I could communicate directly.

I used consent forms and plain language statements were devised for the respondents in which the objectives and broad contours of the research were explained. However, the lack of literacy of many poor farmers meant that I could not obtain their written consent, and therefore had to use verbal consent.

A particular effort was made to ensure that the poor farmers understood the purpose of my interest in interviewing them, as well as the nature of my research, so as to deter false expectations, or any fear or suspicions concerning why I was making inquiries with a range of people within their communities. As I began spending more time in their communities, most of the poor farmers did understand that I was trying to better understand their circumstances, and the ability of ongoing state and donor agency attempts to help improve their lives.

While conducting this research, ethical considerations about the voluntary nature of research were also taken into account, especially when interviewing respondents in vulnerable or unequal relationships, who are not always able to exercise their right of consent. This is because landless and poor farmers are often powerless and thus easily vulnerable to manipulation. Caution was exercised to ensure that these respondents were not under any pressure to take part in the research. A point was made not to approach these respondents through large landowners with whom they have relations of dependency, nor through other prominent powerbrokers within communities such as local politicians or government officials. Instead NGO references were used to gain initial entry to selected villages. While NGOs have some influence in communities around which they are working, the power differentials are not as acute, since many NGOs are themselves aiming to promote empowerment and participation. Therefore, NGOs remained relevant contacts to approach several of the village communities across Sindh and Punjab where research was conducted. Moreover, NGOs were not asked to identify particular research participants, to avoid any sense of compulsion among the respondents. Individual participants were instead recruited to participate in the research after I had developed a personal relationship with the villagers, and different categories of poor farmers, including not only those engaged in sharecropping but also men and women working as seasonal or daily wage agri-labourers.

The small qualitative sample in a limited number of locations cannot claim representativeness of the impact of state and donor policies on poor farmers across the country. An attempt was made, however, to obtain feedback from all distinct categories of respondents (including different types of farmers, state and donor officials involved in varying levels of policy formulation and implementation, and a range of civil society representatives) to explore varied perspectives and dimensions of relevance to the research topic. Overall, I did manage to obtain a sufficiently large, diverse, and complex dataset to explore all relevant facets related to the research outlined in this book.

Structure of this book

Subsequent to the introductory chapter, which has outlined the research aims and questions, as well as the methodological approach, the relevant literature concerning how the Pakistani state and prominent donor agencies like the

World Bank aim to contend with the problems facing poor farmers is reviewed, also highlighting the increasing relevance of peasant movements in this process (Chapter 2). This chapter provides a theoretical overview concerning the major issues which are to be analysed in greater detail. It points out how the state-building process in Pakistan has been influenced by its colonial legacy, and how the Pakistani state has relied on large landowners to consolidate and preserve its hold over power, which in turn has helped the landed rural elite propagate their vested interests through a range of state institutions and policies. Simultaneously, the role of donor agencies in increasingly propagating market-led development strategies is described, as well as how these efforts can reinforce the interests of the landed rural elite. Some of the contentions emerging from state and donor interactions, including the emergence of resistance movements, aiming to alter the rural inequalities, are discussed as well.

Chapter 3 provides an overall profile of poor farmers in the country, and helps situate these poor farmers within the rural political economy in Pakistan. It differentiates between various categories of poor farmers, and highlights province-specific characteristics of farming communities, as well as identifying major challenges confronted by poor farmers.

Subsequent to having provided background details concerning poor farmers, to better contextualize the findings of this research, data gathered during the fieldwork are drawn upon to assess the impacts of ongoing state and donor policies, and the ability of a resistance movement by poor farmers to alter their prevailing situation.

Chapter 4 examines how state institutions and policies deal with poor farmers. Recent legislative attempts, particular features of mainstream political party manifestos, and judicial attempts of relevance to poor farmers are discussed first. Thereafter, problems with a government scheme to distribute state-owned land to landless farmers in Sindh (Benazir Landless *Hari* Scheme) are considered, before highlighting the inability of government departments involved in agricultural development (irrigation, land revenue, agricultural extension, and veterinary services) to specifically address problems facing poor farmers. How state subsidies continue to benefit large landowners, despite the pro-poor rhetoric used to justify their need to donor entities, is illustrated. Finally, some of the emergent biases and practices of relevant government officials concerning poor farmers encountered during the fieldwork are provided.

Subsequently, donor-endorsed approaches to support development processes in Pakistan are considered (Chapter 5). The increasing emphasis on market-based mechanisms to address the challenges of agricultural development and rural poverty is highlighted, with reference to overarching frameworks such as the PRSP for Pakistan. Issues related to using the private sector to promote pro-poor agricultural development are then examined. This is followed by an analysis of an ongoing computerization of land records scheme being funded by the World Bank in Punjab. The ability of NGOs to intermediate between poor farmers and rural land and capital markets is considered as well, before assessing the ability of corporate farming to improve the lives of poor farmers in southern Punjab.

The following chapter then moves on to discuss a recent peasant movement (Anjuman-Muzahereen Punjab, or AMP) which was formed when landless tenants were threatened by eviction from military farms for resisting attempts to replace their traditional farming arrangements with lease-based contracts (Chapter 6). Attention is particularly drawn to the AMP's inability to alter prevalent rural inequalities due to several problems which have come to afflict this resistance movement over time.

Finally, the book presents a cohesive synthesis of the major findings emerging from the simultaneous analysis of state and donor interventions, and the recent resistance attempts by poor farmers, to draw lessons in terms of their combined potential to alter rural inequalities. Reflecting on these research findings, some broader conclusions are drawn for poor farmers in developing countries like Pakistan, where governments are pursuing elite-led development strategies with support from donor agencies like the World Bank (Chapter 7).

Notes

1 The PRSP process has been formulated by the International Monetary Fund (IMF) and World Bank with the aim of encouraging developing countries to develop country-specific poverty reduction strategies to qualify for development loans. The broader PRSP process and the PRSP for Pakistan are discussed in more detail in Chapter 5.

2 The name of this district is not revealed in order to preserve anonymity of research respondents in line with standard academic research ethics requirements. More details concerning anonymity of research are provided in the data sources sub-section of this chapter.

3 While government and donor agencies generally point to the need for 12.5 acres of land to be agriculturally productive, the vast majority of farmers (63 per cent) own 5 acres or fewer (Pakistan Bureau of Statistics, 2011). This book uses the term poor farmers to refer to these farmers with very small amounts of land, as well as several other categories of landless farmers such as small sharecroppers, agricultural labourers, seasonal labourers, and women involved in agriculture. More details concerning these different categories of poor farmers and their comparison with the landed rural elite are provided in Chapter 3.

4 According to the Agricultural Census of Pakistan data, the proportion of farmers with fewer than 5 acres of land has gone up from 58 per cent in 2000 to 64 per cent in 2010.

5 Purposive sampling in qualitative research enables sampling to be conducted on the basis of strategic choices about whom, where and how to conduct research (Given, 2008), which in the case of this thesis was guided by the need to address the three main research questions.

2 State and donor attempts to contend with the problems of poor farmers in Pakistan – a broad overview

In order to understand the challenge of addressing the problems facing poor farmers, it is important to examine both the external and internal dimensions that influence institution building and policy formation processes in Pakistan. This chapter attempts to examine the endogenous factors such as state formation, and the power structure it nurtured and sustained, and the exogenous factors such as donor influence on development policies being pursued in a country like Pakistan, to provide the background required to understand the relevance of both factors in aiding or hindering the effective redress of poor farmers' concerns. Some of the contestations that have emerged from this broader confluence of factors, including resistance attempts by poor farmers, are also highlighted.

The following section begins by examining the historical factors, which led to the dominance of the landed rural elite in Pakistan today. It draws attention to the precedent of colonial land grant schemes, which sought to consolidate the power of the landed elite in lieu of their compliance, and how the post-colonial state adopted similar strategies, which resulted in the increased political influence of the landed rural elite. The colonial and post-colonial precedent of providing land grants to the military is also discussed, to indicate how the most powerful institution in the country also has a vested interest in preserving land ownership inequalities. Subsequent to establishing the significance of the landed rural elite within the broader global production system, this chapter moves on to indicate how specific state and donor interventions have resulted in the consolidation of the landed rural elite and the marginalization of poor farmers. The role of various state institutions in helping to safeguard the interests of the landed elite is considered, by their preventing effective land redistribution from taking place. The failure of state-led and donor supported attempts to address poor farmers' concerns is explained with reference to the 'Green Revolution', for instance, which offered many more opportunities to the landed rural elite rather than poor and landless farmers. Thereafter, the increasing influence of market-led reforms in achieving agricultural development is reviewed, to indicate how these strategies, too, do not offer adequate opportunities to poor farmers. Finally, attention is drawn to resistance movements by peasants, and some of the challenges these movements have confronted, in effectively challenging the dominance of the landed rural elite in Pakistan.

A discussion of these above issues will thus provide a sufficient theoretical background to help to contextualize and assess current state and donor interventions, as well as resistance movement attempts, to address concerns of relevance to poor Pakistani farmers.

Historical evolution of the landed rural elite

In order to appreciate the combined impact of the state and donors on poor farmers it is necessary to consider historical developments in what has now become Pakistan. Too often, prescriptive interventions devised for poor farmers pay little heed to events which took place before Pakistan gained independence in 1947. We, however, look more deeply into changes, which took place in rural areas during the colonial period, the legacy of which shaped the hegemony of the landed rural elite and disempowerment of poor farmers in post-colonial Pakistan.

The Mughal Empire (1526 to 1857) exerted direct control over agricultural land in the Indian subcontinent, and prevented its official aristocracy (*Mansabdars*) from developing land ownership rights under their administration (Ali 1987; Stokes 1978). The British, however, introduced permanent and hereditary rights to lands across rural areas of the Indian subcontinent (Herring 1983). One prominent example in this regard was the colonial land settlement scheme, which accompanied extensive irrigational works to encourage agricultural production in previously uncultivated lands. This took place in the Punjab in particular, but also in Sindh.

The privatization of land had significant consequences for the entire economic structure of the Indian subcontinent. According to Kosambi (1975), the privatization of land gave birth to what he calls 'feudalism from below', compared to the earlier period of 'feudalism from above' in which peasants were largely self-sufficient and were supposed to provide a part of their surplus produce to the king as tribute. The Punjab colonial land settlement schemes were overtly guided by the political considerations of the colonial government seeking to create a class of complicit landowners, consisting of middle-level and an elite peasantry. The British, therefore, turned agrarian tributary societies into more feudal structures which had existed earlier in Europe, by providing permanent entitlements to agricultural land. These land grant schemes facilitated the consolidation of colonial power within rural areas of the Indian subcontinent but they marginalized all other rural classes involved in agricultural production. Agricultural tenants, labourers, and non-cultivating service castes of various types, who had previously held claims to the produce of land, became increasingly dispossessed. In the late nineteenth century, these 'non-agricultural' castes are estimated to have comprised around half of the total population of the Punjab, the most populous province in the country (Gazdar 2011).

Patronage of the peasant hierarchy, as it existed at a given period of time, through provision of private property rights, undermined prospects of social mobility for those categorized as non-agricultural castes. Whereas previously the land management system was flexible enough to adapt to changing socio-economic and political conditions, the administrative machinery of the colonial state consolidated with the position of a minority of the Indian peasantry as the means to legitimizing its own rule. The colonial land administrators also encouraged provision of land grants to incentivize the agricultural castes to work harder, and invest in boosting agricultural production, for providing more raw materials for the manufacturing sector in the United Kingdom (Ali 1987; Gazdar 2011).

The military also benefited immensely from colonial land settlement schemes, including retired military men, such as veterans of the First World War (Ali 1988). Newly cultivable agricultural land was allocated specifically for the purpose of breeding animals, particularly horses to be used by the cavalry. While military personnel were offered land in Sindh as well, the possibility of providing a much greater number of land grants due to the canal settlement schemes in the Punjab led to the British Indian army undertaking a higher proportion of recruitments from the Punjab, leading to the dominance of Punjabis within the military (Ali 1987, 1988). Moreover, the strategic use of land for enhancing the allure of military service enabled the increasing dominance of the military in regions which were to become a part of Pakistan after independence in 1947 (Ali 1987).

Inheriting a particular governance structure due to its colonial legacy had major implications on subsequent policy making in Pakistan as well. Following independence, the founding party, the Muslim League, had to contend with the irreconcilable interests of the landowners and the landless rural masses who had supported the cause of independence lured by promises of economic empowerment within the new nation (Alavi 1972). To placate seemingly contradictory interests, the Muslim League adopted the path of modernization prescribed by the West, which did not seek to redistribute wealth (including agricultural land) but instead aimed to rely on the wealthy to create further wealth. Pursuing such an agenda of economic growth, with the increasing help of donor agencies like the World Bank, did yield growth but its benefits did not really 'trickle down'[1] to the rural poor (Gardezi and Rashid 1983; Jalal 1995).

The notion of the 'overdeveloped' state within post-colonial states like Pakistan also helps explain why institutions of the state such as the army and the bureaucracy are so strong, and why they so readily cooperate with dominant international actors. According to Alavi (1972, 1973), this institutional 'over-development' occurred during colonial rule, which aimed to achieve centre/imperial exploitative interests rather than ensure development of the masses in the periphery/colonized states.

Alavi's analysis of the 'overdeveloped' state in post-colonial societies is, however, unable to explain the contrasting evolution of political processes in the Indian subcontinent (between India and Pakistan). Jalal (1995), for instance, points out that if colonial institutional legacies were so similar, and the underlying class structures only marginally different, then it is difficult to explain why India managed to establish a democratic government while Pakistan continued to experience long episodes of military rule and the continued influence of the military in politics, even during sporadic periods of democratic rule. Besides drawing attention to the differentiated role of the two main political parties (the Congress in India versus the Muslim League in Pakistan), which championed independence for the two countries, Jalal instead highlights the need to consider the strategic and economic consequences of the Indian subcontinent's partition (in 1947) which combined to influence state construction and political processes in the two countries in rather distinct ways. Authoritarianism is considered to have been strengthened in Pakistan due to an interplay of domestic, regional, and international factors, which served to erode prospects of democracy in Pakistan and created a 'political economy of defence' (Jalal 1995: 140). Jalal thus concludes that political power could not have been concentrated in the hands of the civil and military bureaucracy within Pakistan, without the support of the dominant social classes, especially the landed rural elite.

The lack of a robust democratic political culture served the interests of the post-colonial authoritarian state in Pakistan, which also began to rely on local elites (most importantly large landowners) to maintain its control over different parts of the country, bypassing the need for governance based on principles of civic participation or representative democracy. Furthermore, Jalal (1995) argues that military regimes[2] pursued policies of differential economic patronage of these local influentials due to the international development policies of entities like the World Bank and the IMF, since they also endorsed the need for using top-down strategies for achieving economic growth. The economic historian, Imran Ali (1987: 125), interprets the same phenomenon of collusion between the landed rural elite, military regimes, and international agencies within the context of the region's colonial legacy, as follows:

> After 1947, a sequence of military dictatorships has replaced colonial rule to protect the vested interests of the landed. Indeed the older relationship with colonialism has not been disturbed, but has been modified to suit the contingencies of the neo-colonial era. In this brave new world, the native elite, within the lexicon of national status, continues to play a subservient role to external and internationally dominant forces.

Since the allocation of land to the military was also continued by the post-colonial state, the military establishment itself developed a direct interest in preserving the status quo of existing land holdings. According to a defence analyst, Ayesha Siddiqa (2007), the military has come to control around one tenth of all government land in Pakistan. Besides retired military officers being allocated prime agricultural lands, the military is also involved in other agricultural processes such as fertilizer production. Like the landed rural elite, the military also has significant influence over other state institutions, allowing priority access to scare resources including irrigational water, subsidized agricultural inputs, and access to farm-to-market roads.

Donor agencies have not challenged the existing role of the military in agriculture either. Instead, significant amounts of international aid came into Pakistan under the tenure of military regimes in the country. The USA increased economic and military aid to Field Marshal (R.) Ayub Khan during his 11-year reign (1958–69) due to his backing of US geostrategic imperatives during the Cold War. US and multilateral aid to Pakistan also remained significant during the martial law rule of General Zia (1977–89) when Pakistan helped the USA to fight a proxy war against the Soviets in Afghanistan, and then during General Musharraf's reign (1999–2008), when Pakistan partnered with the USA in the 'war against terror'. Given US influence within the governance structures of international financial institutions like the IMF and World Bank (Harvey 2003), US aid was also accompanied by increased support from both these agencies. This international aid helped legitimize the military regimes' role in the governance of the country. Donor policies and programmes accompanying aid to the military regimes did not however directly challenge the vested economic interests of the military, including its control over agricultural land (Ali 2009).

Pakistan's political leadership thus embarked upon safeguarding and perpetuating a narrow set of interests, based on the collusion of the top echelons of the military–bureaucratic establishment, the landed aristocracy, and industrialists. The emergence of the landed rural elite, including the military, is thus rooted in colonial and post-colonial developments, which have enabled them to acquire hegemonic control over rural areas in Pakistan and resist any attempts to alter this status quo. While the underlying factors for the prominence of the local landed elite in the very process of state formation has colonial precedents, the post-colonial state and donor agencies have continued to preserve the vested interests of the landed elite, including the military.

Consolidation of the landed rural elite and marginalization of poor farmers

Besides historical factors, which helped forge an underlying convergence of interests between the landed elite, the post-colonial state, and dominant international powers, specific state and donor interventions have further strengthened the landed rural elite while undermining the position of poor farmers in Pakistan.

At the time when the Pakistani state was created, mercantilist[3] and state-centred notions of development were still influential among nascent international development institutions. Donor agencies like the World Bank and the IMF also promoted centralist models of development when they began lending to Pakistan in the early 1950s. Classes were assigned importance according to the interests of the state. The landed aristocracy had the responsibility of making the peasantry work, so that the surplus extracted from them could be utilized in domestic consumption by the gentry and for the production of raw materials for foreign trade. Traders and manufacturers were considered the backbone of the mercantilist political economy. The poverty of labourers was considered necessary to enable wealth creation through expropriating their produce (Gardezi and Rashid 1983; Hasan 1997).

However, when mercantilist ideas were adopted by the post-colonial Pakistani state, they did not lead to the expected successful transition to industrial capitalism. Perhaps this is because the Pakistani state was not built independently, as in classical mercantilism, but instead Pakistani mercantilists remained linked with imperial powers even after independence. Thus, the very notion of protectionism was defined not to serve broader national interests, but rather to protect a much narrower set of elite interests. The post-colonial authoritarian state, which emerged in Pakistan, thus continued to preserve the vested interests of the local elite (including large landowners), while pursuing imported notions of achieving growth, even as the country entered a new stage of economic planning under the economic philosophy of laissez-faire (Gardezi and Rashid 1983).

The introduction of 'Green Revolution' technologies in the 1960s provided the largest landholders in the country an opportunity to become vital partners in the elusive goal of economic growth, without, however, disturbing a status quo marked by large social and economic inequalities (Hussain *et al.* 2003). In Pakistan, the 'Green Revolution' was introduced to the agricultural sector with the explicit objective of eradicating rural poverty without undertaking radical land reforms (Alavi 1973). Policy makers resolved to address the problem of lacklustre economic growth in rural areas by emphasizing the need for investing in high-yielding varieties of food grain and focusing on the rich, who had the money to buy the necessary agricultural inputs, to generate high growth rates (Alavi 1973; Gardezi and Rashid 1983).

However, it was mainly large landholders who benefited from these developments rather than those with small holdings. The reason was that large landowners made the greatest progress in the direction of farm mechanization. Smaller landowners could not keep pace with the high cost requirements of intensive farming, and many of them began to lease their land out to capitalist producers with the economic resources to do so. Due to mechanization, such as the increasing use of tractors, sharecroppers also began to be evicted from their holdings, notwithstanding their supposed legal security of tenure (Herring and Kennedy 1979).

A major reason why the 'Green Revolution' and subsequent state- and donor-endorsed attempts to undertake agricultural development largely served to benefit large landowners was due to the lack of effective land reforms in the country. Land reform legislation was first enacted in Pakistan under the martial law regime of Field Marshal (R.) Ayub Khan in 1959, and another land reform act was passed in 1972 under the democratic government of Zulfiquar Ali

Bhutto. These land reforms imposed ownership ceilings, beyond which agricultural land holding was to be subjected to redistribution.[4] However, ceilings on the amount of land were placed on land owned by individuals rather than agricultural families (Herring 1979). Most large landowners simply transferred land ownership titles to their own family members in order to circumvent the planned land redistribution (Gazdar 2011; Herring 1979; Hussain 1984).

The very parameters of the land reform debate within the country are criticized for being myopic due to their primary focus on addressing the plight of landless cultivators, instead of also raising the issue of labouring classes, including seasonal and daily wage earners (Gazdar 2011; Herring 1979; Hussain 1984). Land reforms in Pakistan also did not question the social hierarchies that had been formalized under colonial rule. For instance, the 1959 Land Reforms Commission, staffed by bureaucrats, did not deal with the problem of the sizable proportion of agricultural labourers in rural areas within the land reform process, nor did it devise any other specific plans for their empowerment, such as introduction of minimum wages (Gazdar 2011).

The initial land reform attempts under the Provincial Tenancy Act of 1950 granted landless tenants some modest concessions, but these proposals faced strong resistance in the provinces of Punjab and Sindh by the landed rural elite (Brohi 2010). Revised legislation under the Land Reform Act of 1972 also attempted to regulate sharecropping rents for landless tenants, by preventing the landlord from obtaining more than 40 per cent of the gross agricultural yield produced by their sharecroppers, but these measures were not implemented effectively either (Herring 1983). The Pakistani state, or major donors like the World Bank, did not show the willingness or resolve to intervene and regulate the terms of contracts between large landowners and their landless tenants. As a result, the 1972 census confirmed the widespread violation of the earlier tenancy laws, with over 80 per cent of sharecroppers still giving 50 per cent of their yield in the form of rent to landowners, instead of 40 per cent as stipulated by the Land Reform Act of 1972 (Herring 1983).

A mere 1 per cent of landless tenants and small landowners are estimated to have directly benefited from the 1972 reforms (Chaudhry 1974). Conversely, attempts to regulate tenurial relations instead led to tenant evictions by larger landowners who began to instead rely more heavily on daily waged and seasonal labourers, since there was not (and still is not) legislation offering these disempowered agricultural workers any form of protection or benefits.

The former Prime Minister of Pakistan, the late Zulfiqar Ali Bhutto, did also promulgate a National Charter for Cultivators in 1977. This charter mandated that all cultivable state lands not being used for a public purpose be given to landless cultivators with less than a subsistence holding. Bhutto also tried to change the land-revenue taxation mechanism to make it more progressive (Herring 1983). The overthrow of the Bhutto government in 1977, and the subsequent imposition of martial law, prevented these laws from taking effect. Yet, even if the Bhutto government had not been overthrown, the fact that many large landowners managed to become Bhutto-backed candidates in the elections of 1977 implied that further land reform attempts might very well have been subverted from within his government. In subsequent years, many of the members of the provincial and national assembly have continued to belong to prominent landowning families (Ali 2010a). Hence, chances of further land reform legislation being passed and implemented remain unlikely.

Even if land reforms were to occur in Pakistan, it would be necessary to pay attention to what supplemental measures are required, besides the provision of land, to ensure the success of poor farmers. International experience indicates that while redistributive land reforms have played a crucially important role in the rapid growth of East Asian countries such as South Korea, Taiwan, and Malaysia (Herring 1983). The results of land reforms in many other parts of Asia, Africa, and South America were much less positive. In Nicaragua, Honduras, and El Salvador, for example, where agricultural production and standards of living initially increased among cultivators, but then the lack of easily available agricultural loans and ineffective agricultural extension services, coupled with broader economic problems such as increasing inflation, began undermining the initial redistributive gains of land reforms (Pereira 2005). This lacklustre result of land reforms in these latter countries placed singular focus on redistribution of land, while neglecting the supplemental policies required for land redistribution to become economically productive. The establishment of adequate support mechanisms for poorer farmers is a vital aspect of land reforms, but it does not get the required attention within land reform programmes launched in many developing countries, including Pakistan (Ali 2010a, 2010b).

Providing an enabling institutional structure, which is supportive of poor farmers, remains a difficult proposition in the case of Pakistan. Instead, different institutions of the state actively seek to preserve the interests of large landowners. Siddiqa (2007) points out how even the judiciary in the country has played a detrimental role in terms of safeguarding the interests of poor farmers. The Federal Shariat (Islamic) Court[5] ruled in 1980 that land reforms were un-Islamic and, therefore, unconstitutional. The Supreme Court upheld this decision in 1989 (Siddiqa 2007). The courts' declaration that the mandatory acquisition of private land by the use of state force was against the principles of Islamic law has put the issue of land reforms on the backburner in the country, despite lingering rural disparities and deprivation.

It has been argued that 'feudalism'[6] is on the decline in Pakistan in the twenty-first century, and that phenomena like land fragmentation,[7] the rise of urban or religion-based political parties, and a new class of industrialists and commercial real estate barons have been encroaching on feudal economic power (Lancaster 2003). However, such assessments can be easily refuted.

The landed rural elite has, in fact, sought to diversify itself, investing in businesses such as textile mills and preparing its offspring for professional careers, including in the army, the bureaucracy, and even donor agencies. This diversification of influence enables the landed rural elite to preserve the status quo of land ownership inequalities based on broad-ranging institutional support.

Donor agencies like the World Bank, for instance, have also played an increasingly important role in reinforcing the position of the landed rural elite. On the one hand, this has occurred through an emphasis on capital-intensive agricultural development interventions, which have mainly been to the benefit of larger farmers who have the required resources and leverage with state institutions. These farmers are therefore able to take advantage of available schemes intended to facilitate agricultural development. On the other hand, donors like the World Bank have failed to address the need for redistributive land reforms in Pakistan. Instead, donor agencies have developed top-down technocratic solutions, such as emphasizing the need for the provision of different types of subsidies or rural credit, which have little value in terms of addressing the range of inequalities, which continue to marginalize poor farmers (Zaidi 2012a). By ignoring critical issues such as the need for encouraging equity in terms of land ownership, donor agencies remain unable to prevent the myriad abuses of power emanating from the lack of land ownership by vast segments of the rural populace.

Even though capitalism has pervaded rural Pakistan, capitalist processes often operate through traditional (pre-capitalist) forms of social organization and governance norms (Malik 2008).[8] Land not only provides economic collateral, but also social collateral, which in turn can be leveraged to achieve recognition within kin, tribe, or caste-based networks. Such networks are, in turn, important resources needed to ensure individual access to markets and other public resources (Chaudhry *et al.* 2006; Cheema *et al.* 2009; Gazdar 2005).

Thus, many of the problems that exist in rural Pakistan, such as the exploitation of the rural poor by a small number of landed elites, continue to emanate from uneven landholding patterns, since land ownership is not simply a matter of economic power; control and ownership of land allows a disproportionate control over a range of other (economic, socio-cultural, and political) assets (Hussain 2005, 2008). The Pakistani state's patronage of the landed rural elite not only helps them capture state resources, but to also avert reforming major causes of prevailing rural inequalities, particularly the uneven patterns of land ownership in the country. Since donor agencies have not directly contended with altering power relations associated with landownership in rural areas, they have further enabled the capture of state-led and donor supported agricultural development schemes by large landowners. Resultantly, elite-led agricultural development, subsidized by the state, and supported by donors, lacks the capacity and the inclination to improve the lives of the multitudes of poor farmers in the country.

The state, markets, and poor farmers

After the failure of donor support for state-led measures such as the 'Green Revolution' to achieve goals like economic growth or rural poverty alleviation, donor agencies like the World Bank began adopting neo-liberal policy prescriptions emphasizing the role of market forces in achieving development. From the 1980s onwards, emphasis was placed on minimizing public expenditures within developing countries, and instead promoting market-based interventions, including in the agricultural sector (Besley and Burgess 2000; Khan 1999). Almost three decades of structural adjustment programs, instituted as part of IMF and World Bank loan packages to developing countries, resulted in diminishing subsidies for farming inputs like fertilizers and seeds, which poorer farmers, including landless cultivators, now also increasingly rely on. Implementing structural adjustment programs also encouraged agricultural production in the developing world to shift focus from food production for domestic (local) consumption to cash-crop production to obtain export earnings (Skogly 2001).

Under increasing pressure from the emergent global production system, it is not targeted state interventions or donor support, but rather desperate self-exploitative measures which have ensured the survival of poor farming households, such as engaging in child labour and placing an increasing burden of agricultural production onto women (Hussein *et al.* 2004; Thorner *et al.* 1986). In Pakistan, poor women in rural areas also have a major share in the country's agricultural production, yet their work is not adequately remunerated, and they hardly hold any productive assets, especially land (Brohi 2010; SDPI 2008).

The World Bank's Doing Business rankings, formulated in 2002, began scoring countries according to their ability to foster an enabling environment for 'doing business'. Research on the impacts of this ranking point towards a resulting push for liberal reforms within the developing world, such as adoption of investor-friendly regulations and suppression of trade barriers, as well as introduction of intellectual and structural framework for further deregulation of the agricultural sector, encouraging, for instance, the growing trend of large-scale land grabs in developing countries. African countries like Liberia and Sierra Leone, for instance, implemented dozens of investment reforms to attract foreign direct investments, allowing for instance guarantees against unfair expropriation of investors, and ensuring their ability to repatriate capital and profits. Within the agricultural sector itself, these 'business-friendly' reforms have encouraged palm oil, sugar, and rubber giants such as the British Equatorial Palm Oil, Addax Bioenergy from Switzerland, Quifel Natural Resources from Portugal, which have acquired millions of acres in just a few years, taking away farms, resources, and livelihoods from multitudes of poor farmers (Martin-Prével 2014a; World Bank 2013). Pakistan too has put in place legislative mechanisms to facilitate corporate farming as well as leasing of state land by international agri-business concerns, as will be discussed in more detail in Chapter 4.

Moreover, in 2012, the G8[9] asked the World Bank to develop a more specific index for Doing Business in the agricultural sector. With funding from the US, UK, and other European governments, the Benchmarking the Business of Agriculture (BBA) project thus emerged in 2013.The new indicators' aim is create a ranking based on efforts to allow a more modern, and presumably a more productive, agriculture and agri-business sector to emerge. The stated goals of the Doing Business, and now the BBA, are to inform policy makers about regulations that favour business in their countries (World Bank 2014). In effect, however, the wide promotion of such annual indexes among other international donors and foreign investors, the World Bank is criticized for applying a simplistic logic of comparison and confrontation between nations. Such indexes praise supposed 'good performers' and stigmatizes low-scoring countries, compelling a further race to deregulation, especially among cash-starved governments in developing countries as they compete with each other to attract international investment by appearing more 'business friendly' (Martin-Prével 2014a, 2014b).

The BBA, like the earlier World Bank's Doing Business index, also places nations on a linear development path, encouraging them to adopt a one-size-fits-all model, which compels them to pursue agricultural transition through input- and capital-intensive agriculture. The BBA notably is even more specific in the policies it endorses within the agricultural sector, which include, for example, lower tariffs on commercial seeds and fertilizers (World Bank 2014). The BBA does not however pay due attention to the need for rehabilitation and improvement of traditional seed production, plant breeding, and agro-ecological techniques can increase productivity while also enabling self-reliance and independence of poor farmers from ever-fluctuating international agricultural markets (Martin-Prével 2014a, 2014b). Thus, while there are evident opportunities for furthering private agri-business concerns by encouraging this capital-intensive ranking, the World Bank has not formulated specific criteria whereby poorer farmers could benefit from the benchmarking of the agricultural sector.

The propagation of market-led and neo-liberal reforms[10] within the agricultural sector also explains why inequality in land ownership has become a dominant feature across rural areas in developing countries (Birner and Resnick 2010; Borras 2008). Neo-liberal restructuring is criticized for having expanded the scope of already established inequitable land ownership and by increased commodification of agricultural production relations, and exposure of an even greater number of people to the market imperative (Borras and Franco 2010; Byres and Bernstein 2001; Toor 2010). This is due to the World Bank's use of market-based principles to contend with the issue of unequal land ownership, which often entailed support for market-based redistribution schemes, and emphasizing the need for land titling and registries to create more effective land markets. The World Bank has supported such programmes in different developing countries, including South Africa, Brazil, and India. Yet such market-assisted land reforms are problematic, since they place a heavy financial burden on poor farmers to repay expensive loans for the land they have purchased from large landowners, while struggling to ensure household survival without adequate institutional support (Rossett 2001).

Despite lacklustre results, the World Bank has also been encouraging Pakistan to use similar approaches to address the problems of agricultural development and rural poverty alleviation. While acknowledging that prevailing land tenure arrangements in the country are too skewed and inequitable to increase agricultural production, the World Bank reiterates the need for ensuring the transparency of land records, which it presumes will give an impetus to investments in agricultural production and also to the process of increasing land transactions needed to achieve efficient land use (Mansuri and Jacoby 2006). Whether focusing on ensuring secure land tenures, or facilitating land transfers, are adequate measures to address problems like rural poverty in the country is debatable, however. Creating more efficient land markets within the prevailing rural setup in Pakistan may merely encourage large landowners to further displace sharecroppers, and lease out their lands for capital-intensive farming by local or foreign agri-businesses. Yet these are risks that the World Bank itself does not adequately address.

Although the World Bank's 'World Development Report: Agriculture for Development' (2008), concedes that smallholders (poor farmers) often remain the most efficient producers, however, it then points out that these poor farmers cannot capture economies of scale in production. Hence, it suggests labour-intensive commercial farming as the key instrument to reducing rural poverty. Yet, the ability of commercial farming, led by agri-business firms in particular, utilizing labour-intensive farming methods, and offering well-remunerated opportunities to its labourers, is doubtable.

A combination of state and donor policy priorities in countries like Pakistan has, nonetheless, resulted in an increasing convergence of transnational and domestic capital to influence the process of agricultural production (Toor 2010). But the instigation of privatization and deregulation policies by the World Bank and IMF, interlinked with the WTO trade rules, are having difficulties in delivering agricultural development and rural poverty alleviation goals. In fact, the outcome is often the opposite of neo-liberal reformers' predictions. It has led to the Philippines, for example, from being a net agricultural exporting to a net agricultural importing country (Borras 2006). Liberalization of the agricultural sector in the Philippines has largely reinforced pre-existing agrarian structures dominated by national and transnational elites, without addressing the persistence of poverty and growing inequality in the country. Similar conclusions would not be out of place in the case of Pakistan, where half the rural population has experienced varying levels of poverty over the past decade, despite reasonable periods of agricultural growth (PIDE 2012).[11]

Over the past decade, international donor agencies have conceded to the need for increasing country ownership of internationally funded development efforts and ensuring participatory approaches in development planning to enhance the effectiveness of international aid (Stiglitz 2002). Yet, the extent to which policy frameworks, such as the Poverty Reduction Strategy Papers (PRSP),[12] have been able to promote participatory policy making within developing countries remains questionable (Forster and Schnell 2003). The PRSP process in Pakistan is spearheaded by the Ministry of Finance, yet the final approval of PRSP strategies remain conditional on IMF and World Bank approval.

It is, therefore, not surprising that the government of Pakistan's PRSP documents make explicit references to corporate agriculture, and increasing agricultural productivity by exploiting economies of consolidation and scale, rather than expressing the aspirations, or addressing the concerns, of poor farmers, who comprise the majority of the rural workforce in the country (Government of Pakistan 2003). These agricultural policies are commonly found in a range of other developing countries' PRSP documents as well, which are World Bank and IMF loan recipients, given the World Bank's preference for adopting capital-intensive farming as one of the major strategies to boost agricultural yields (McMichael 2009).

The World Bank envisions a limited role for the state, primarily in terms of correcting market failures for the greater inclusion of smallholders and rural workers in agricultural production, or enhancing their access to micro-finance to help them invest in increasing their agricultural productivity (World Bank 2007). The provision of micro-finance alone, however, does not provide poor farmers leverage enough to challenge the power of the landed rural elite, or to overcome the varied institutional biases confronted by them, nor does it allow them to spearhead donor-backed market-based strategies for enhancing agricultural productivity.

Given the fact that the Pakistani state itself is held captive to preserving the vested interests of the landed rural elite, it is hard to see how the state could play an effective, enabling role to include poor farmers within the donor-prescribed market-led agricultural growth strategies. Conversely, it is the agricultural estates of the larger landowners that show characteristics of modern capitalistic farming, such as the extensive use of wage labour, and the use of modern inputs and machinery. Thus, larger landlords have a much greater chance of benefiting from market-based principles of economic liberalization than poorer, marginalized farmers (Gazdar 2009). This, in part, also explains the willingness of the landed rural elite within Pakistan to readily agree to the adoption of market-based strategies promoted by entities like the World Bank, including capital-intensive corporate farming. The ability of World Bank-supported development strategies, which promise growth within, rather than outside, the existing social relations, to be able to benefit poor farmers, thus remains problematic.

Emerging contestations and concerns

The opening up of agricultural markets across the developing world has had an immense impact on the dynamics of agricultural production between and within different countries of the world, often to the detriment of poor farmers (Borras 2008; Borras *et al.* 2008; Borras and Franco 2010). As a result, there is a growing emphasis on the need for improved cooperation between donors, the state, and civil society to help address the problems of poor farmers (Borras 2008). Yet, a mutually enhancing relationship between many mainstream civil society organizations, the state, and donor organizations has been criticized as being underpinned by a neo-liberal assumption, rather than an engagement with wider debates about the politics of development (Ferguson and Lohmann 1994).

Within an overarching imperative of minimizing the role of the state to lessen public expenditures, entities like the World Bank have often encouraged the private sector and civil society organizations to take on the responsibility of improving social service delivery (health, education, sanitation, etc.). While the notion of civil society itself is much larger, non-government organizations (NGOs) occupy a prominent position as representatives of civil society, in large part due to the key role NGOs now play in donor- and government funded- development efforts (Mercer 2002). NGOs are at the forefront of not only implementing several donor-supported programmes, but also facilitating market-based solutions such as the use of micro-finance to address structural problems such as poverty (Ghaus-Pasha and Iqbal 2003). Such donor-supported NGO interventions are recasting political questions about land, resources, or wages as technical problems requiring development expertise, and are thus deemed controversial. Ferguson terms such processes of development to be an 'anti-politics machine' which, in emphasizing goals of delivering narrow development outcomes, turns attention away from the need for achieving empowerment through the process of political mobilization (Ferguson and Lohmann 1994). From such a perspective, the work of many NGOs may be viewed as undermining the impetus for meaningful change. The creation of quasi-autonomous non-governmental organizations (QUANGOs) is looked upon with similar suspicion (Mercer 2002).

Yet, despite concerns that governments and donor organizations work only with civil society organizations that adopt complicit positions in terms of their organization values and outlook, there is also evidence of civil society efforts to resist prominent state and donor interventions, and to support the mobilization of marginalized groups, including those comprised of poor farmers.

There has in fact been a noticeable growth of peasant and indigenous movements around the developing world over the past two decades. The Zapatistas in Mexico, the Landless Rural Workers' Movement in Brazil, and the anti-Nirmada Dam movement in India are prominent examples in this regard (Borras *et al.* 2008). Another recent resistance movement by landless farmers in Pakistan has also been lauded for taking a bold stance, resisting the eviction of landless farmers from military farms in rural Punjab (Mumtaz and Mumtaz 2012; Zaidi 2012b). Present-day peasant movements and networks are quite diverse. They are highly heterogeneous in terms of class, ideologies, and in their scope of action. The peasant movement in the Punjab, for instance, is not pitted against large landowners, but against state-owned agricultural lands controlled by the military. One reason why the resistance movement against the military farms in Punjab is considered so important is due to its having brought together different categories of landless farmers, transcending caste, religion, and gender differentiations that often become impediments to collective action (Akhtar 2006). Yet, it is the very heterogeneity of this resistance movement, which has subsequently caused fragmentation, and diluted its impact on altering the status quo of land ownership in Pakistan (discussed at length in Chapter 6).

The mere participation of poor farmers in land reform networks is often not sufficient to redress unequal power relations, and problems of social exclusion for formerly landless cultivators are seen to continue, even after they gain access to land (Lindemann 2010). In order to improve the lives of poor farmers, it is vital that peasant movements forge a sense of collective identity and also create effective linkages with broader social movements within and across developing countries. The need for creating such linkages provides civil society organizations working on land rights and social justice issues to play a significant networking role. Among the most visible of transnational peasant movements is Vía Campesina,[13] which illustrates the possibility for rural citizens of different countries to collectively begin asserting their rights to define what land reforms mean to them.

Collectively, peasant movements are considered to have accomplished a level of influence that can no longer be ignored by states, or mainstream donor agencies like the World Bank (Borras *et al.* 2008). However, analysis of transnational agrarian movements has only just begun to recognize the complicated historical origins and the delicate political balancing acts that necessarily characterize any effort to construct cross-border alliances linking highly heterogeneous organizations, social classes, ethnicities, political viewpoints, and regions (Borras 2008; Borras *et al.* 2008; Borras and Franco 2010). How transnational agrarian movements can provide greater leverage to groups of poor farmers in contending with local power hierarchies, which are unique to different settings, is still not well understood. Unless greater attention is paid to how local peasant movements can be linked effectively with broader coalitions, the generic rhetoric of empowerment associated with such larger international movements, and the on-ground realities facing poor peasants (including landless cultivators) within their local communities, may remain unbridgeable. This is a major issue confronting many prominent peasant movements across the developing world today, including in Pakistan.

Conclusions

The survey of literature in this chapter has provided a theoretical backdrop to understanding the persistent dominance of the landed rural elite in Pakistan. While colonial developments, which occurred in rural areas that now comprise

Pakistan, played an important role in the evolution of the landed rural elite, the patronage of landowners was continued by the post-colonial state. The growing importance of the landed elite in state formation processes further explains their ability to safeguard their vested interests with the support of a range of state institutions. Furthermore, the colonial precedent of providing military control over agricultural lands was also continued by the post-colonial state, and reinforced by aid to military regimes in the country. However, both the colonial and subsequent post-colonial governments in Pakistan did not pay much attention to empowering poor farmers, who primarily remained involved in sharecropping or worked as daily wage labourers.

Both state and donor attempts to boost agricultural production have enabled the landed rural elite to benefit disproportionately from agricultural development policies in comparison to poorer farmers. Donor-supported state-led strategies for agricultural growth led to the 'Green Revolution', which promoted an emphasis on mechanization and subsidized larger landowners in the effort to boost yields, rather than focusing on improving the productivity of poor farmers. The emphasis on elite-led growth strategies was not deterred under the influence of neo-liberalism either, when the World Bank increasingly began emphasizing the need for effective market mechanisms rather than state interventions to achieve both agricultural growth and rural poverty alleviation. Neglecting the need for redistributive reforms, however, has been further marginalizing poor farmers despite tokenistic attempts such as the provision of micro-finance schemes, which can do little to address the prevailing power imbalances, which remain tilted in favour of the landed rural elite.

While the rhetoric may have changed over the past decade or so, with an emphasis on less state-centric and more participatory growth, the opportunities offered by emergent market-led agricultural development approaches also appear to be more relevant to larger landowners rather than poor farmers. The continued emphasis on boosting agricultural yields through further liberalization of agricultural markets and emphasis on capital-intensive agricultural development within the Poverty Reduction Strategy Papers for Pakistan, also offers scant opportunity to benefit marginalized communities (such as poor farmers) within the country.

The above analysis indicates that large landowners, the Pakistani state, and donor agencies such as the World Bank, can in fact be described as having certain basic interests in common; such as the preservation of the existing social order, based upon the institution of private property. Yet, despite the evident prominence of the landed rural elite within Pakistan's political economy, and the potential convergence between their interests, those of the state, and donor agencies, the relationships between these different stakeholders are not seamless. Increasing complications have occurred despite these common interests over the past three decades, such as pressure by donor agencies to curb government spending, including subsidies, which benefit large landowners, and increasing revenues by taxing the agricultural sector. The ways in which different state institutions have chosen to implement or ignore donor policies, and how this interaction helps preserve the vested interests of large landowners, is, however, still unfolding.

On the other hand, the role of peasant movements in challenging state and donor policies has begun receiving increasing attention. Yet peasant movements have a long history in parts of the Indian subcontinent, now comprising Pakistan. There is also evidence of a recent resistance movement led by landless farmers, which has sought to challenge the military's control over agricultural lands in the country. However, there are numerous challenges that are diminishing the possibility of such a significant achievement. What these undermining factors are, and why they remain unable to effectively resist existing rural inequalities, are also issues which remain of vital significance.

In view of the theoretical context developed within this chapter, the following chapters of this book now examine a range of relevant state and donor interventions, which are ongoing in Pakistan, to demonstrate how they too continue to benefit larger landowners rather than poorer farmers. Thereafter, the existing problems, which are hindering resistance by landless farmers in rural Punjab to effectively challenge the prevailing marginalization of poor and landless farmers in Pakistan are discussed. Prior to presenting details of state and donor interventions, and resistance to them, it is imperative to first provide detailed information situating poor farmers within their broader surroundings, and highlight the major challenges they presently confront. The next chapter, therefore, provides an overall profile of poor farmers and their situation in Pakistan.

Notes

1 The 'trickle down' approach to development presumed that state or elite-led capital intensive models of investment would stimulate growth and in turn enable prosperity to flow from the top to the lower levels of society.

2 Since independence, Pakistan has experienced long periods of military rule; from 1958 to 1971, from 1977 to 1988, and from 1999 to 2008.

3 Mercantilism itself developed during the decay of feudalism in Europe with the aim of unifying and increasing the power, and especially the monetary wealth, of a nation by strict governmental regulation of the entire national economy, usually through policies designed to achieve a favourable balance of trade. The central role of the government in managing development processes exerted influence within international agencies like the IMF and World Bank, within the post-Second World War context.

4 The land ownership ceilings have varied under the different land reform attempts. In 1972, however, Bhutto had reduced land ownership ceilings to 300 acres of un-irrigated land and 150 acres for irrigated lands.

5 The Federal Shariat Court of Pakistan consists of eight Muslim judges, including the Chief Justice of the Supreme Court. The FSC, on its own motion or through a petition by a citizen or a government (federal or provincial), has the power to examine and determine as to whether or not a certain provision of law is repugnant to the injunctions of Islam, and the government is required to take necessary steps to amend the law under question to bring it in conformity with the injunctions of Islam.

6 While the term feudalism has its origins in describing the political and economic system of Europe from the ninth century until about the fifteenth century, based on

the holding of all land in a fief, today the term is frequently used in developing countries to describe conditions of servitude, as well as ordered oppression, on the basis of land holdings. This thesis however does not use this term unless in reference to its usage by others. Instead, those with significant landholdings are described as large landowners.

7　The process of inheritance over generations naturally leads to a decrease in land holdings over generations. While some analysts point to land fragmentation being a major challenge, others refute this claim pointing out that land is still sufficiently concentrated to justify the need for its redistribution.

8　Some of these forms of social organization include kinship-based groups, for instance, and informal governance mechanisms include informal arbitration mechanisms such as *panchayats*.

9　Formed in the 1970s the G8 is a grouping of the most industrialized countries including the United States, Britain, Germany, Japan and Italy, Canada, and Russia. The President of the European Commission, who represents the European Union, also attends the G8 annual summits.

10　'Neo-liberalism' is a set of economic policies which support economic liberalization, free trade, deregulation, privatization, and decreasing the role of the public sector. The promotion of neo-liberalism in developing countries which borrow money from the World Bank and IMF is commonly referred to as the 'Washington Consensus'.

11　The Government's Pakistan Economic Survey report for 2009, for instance, claimed that the agricultural sector had been growing at an average annual rate of 4.1 per cent since 2002–3 (Government of Pakistan, 2009).

12　The PRSP process was formulated and launched by the IMF and the World Bank at their annual meeting in Washington, DC. It required developing countries to formulate their own strategies for achieving development, on the basis of which concessional loans would be provided to them. Pakistan thus far has produced two PRSP documents.

13　La Campesina, or the international peasants' movement, claims that it brings together millions of poor farmers through 150 local and 70 national organizations from across 70 countries, takes up their concerns, and it champions the principles of sustainable agricultural development against corporate-driven agriculture.

3 Situating poor farmers in Pakistan's rural political economy

Since this book is primarily concerned with the ability of state and donor development efforts to improving the lives of poor farmers, it is necessary to situate Pakistani farmers, especially poor farmers, within the broader political economy of the country. This chapter, therefore, begins by providing a broader overview of farming communities in rural Pakistan, the significance of agriculture, as well as landownership patterns in the country. The next subsection differentiates between different categories of farmers, also highlighting how poor farmers themselves are not a homogeneous group. Next, specific developments concerning farming communities within each of the two provinces (Sindh and Punjab) where research was conducted is provided. Some of the prevailing challenges facing poor farmers, including women involved in agriculture, are discussed, prior to concluding by emphasizing the need for effective efforts to address the concerns of poor farmers in the country.

Historical perspective on Pakistani farmers

More than two-thirds of over 180 million Pakistanis live in rural areas, and 40 per cent of country's total labor force is involved in agricultural activities (World Bank 2014). Since most of the poor people in rural areas do not own cultivable land, this in turn leads to various forms of exploitation. Besides drawing attention to the problem of uneven rural landownership, this chapter will indicate how different categories of farmers fare within rural areas of the country, and ascertain the reasons for the variance in their circumstances. Reference to a historical perspective, will also help to understand the present day position of different categories of farmers within the prevailing rural political economy of the country.

The agricultural lands that are now located in Pakistan have experienced centuries of foreign rule by Muslim rulers of Central Asian origin, including the Mughals, and then by the British from the later part of the nineteenth century until 1947. The theory of the Asiatic mode of production, devised by Marx in the 1850s, viewed Asiatic societies, including that of the Indian sub-continent, as comprising largely undifferentiated and self-sustaining villages controlled by despotic rulers, themselves residing in cities and expropriating rural surplus to finance their lavish lifestyles (Mitrany 1952). Conversely, scholars from the Indian Subcontinent, like Kosambi (1975) differentiate between two distinct stages of feudalism in pre-colonial India; which were described by the terms (a) feudalism from above, and (b) feudalism from below. The first term aims to describe the state of affairs wherein powerful rulers levied a tribute from subordinates, but allowed them to exert direct control over smaller territories. These subordinate rulers, such as tribal chiefs, were said to have ruled their territories by direct administration, without the intermediary of a landowning class. A class of landowners gradually began developing within villages, to play an intermediary role between the state and the peasantry.

Nonetheless, both the Muslim and Hindu conception of agricultural land-ownership remained embedded in the principle that the field is the property of the man who first brings it under cultivation. However, increasing extractive pressure applied by the Mughal Empire during its declining years, resulted in armed opposition to the Mughal state. In this process the older regional elite was destroyed by peasant war-bands from among whom emerged new fief-doms, which were acknowledged by the British in lieu of their willingness to side with the expanding British Empire. This new dominant class of farmers considered itself superior to the host of service and labouring castes (generically referred to as *kamais* or menial labourers), which constituted a majority of the rural population (Ali 1988; Stokes 1978).

As British rule over the Indian subcontinent consolidated, colonial administrators began experimenting with different types of property ownership and revenue generation systems. The permanent settlement instituted by Lord Cornwallis in 1793, when the now Indian states of Bengal, Bihar, and Orissa were ceded to the East India Company, turned peasants into tenants and the estate administrators and tax collectors (*zamindars*) into landowners, responsible for paying fixed rents to the colonial administration. East India Company officials had assumed that if *zamindars* became like the British landlord class, they too would adopt a capitalist ethic and help boost agricultural productivity and trade. These *zamindars* however sub-leased their revenue tracts to a group of intermediaries, who ruthlessly exploited tenants, to extract maximum profits for themselves and the *zamindars*, rather than being concerned with productive administration of their estates. As the British Empire consolidated its hold on the Indian subcontinent, other revenue collection systems were also implemented in southern India for instance, where the *ryotwari* system, in which the government settled land revenue directly with the peasant cultivators through a system of temporary settlements, which classified agricultural fields according to soil type and produce, with average rent rates fixed for the period of the settlement (Stokes 1989). However, these varying revenue collection arrangements were guided by the underlying imperative of securing support for a British presence in the subcontinent and obtaining uninterrupted revenues initially for the East India Company, and, thereafter, directly for the British Empire.[1]

The alliance between dominant agrarian groups and the colonial state was strengthened through extensive land settlement schemes on new canal irrigated tracts from the 1880s onwards. In these agricultural colonization schemes of the British Empire, the grantees of land once again came primarily from the 'agricultural castes', or were already landholders. Lands were allotted to civil employees and military personnel, which were mostly, but not entirely, from landed castes. As mentioned in the preceding chapter, poor landless farmers and agricultural labourers, however, did not benefit from colonial land grants (Ali 1987, 1988).

Caste, a feature of the Hindu tradition, thus continued to be a prominent aspect of village life despite the spread of Islam, which theoretically does not support the idea of a caste system, in the Indian subcontinent. Even under the Mughal Empire, the subjection of the menial proletariat by the caste-based peasantry persisted uninterruptedly (Habib 1983). The colonial state's decision to exclude non-agrarian kinship groups and castes from being able to access landownership, also endorsed the prevailing status quo of landownership rights (Ali 1988; Cheema *et al.* 2009).

Despite the abolition of colonial legislation preventing non-agrarian castes from acquiring land at the time of Pakistan's independence in 1947, landless farmers such as sharecroppers and agricultural labourers, remain largely excluded from both land and tenancy markets (Cheema *et al.* 2009; Gazdar 2005). Thus, while there is a wide category of farmers involved in agriculture in modern day Pakistan, a majority of them remain marginalized and disempowered due to their lack of direct control over sufficient land.

Farmers in post-independent Pakistan

Ownership of cultivable land in Pakistan remains uneven, with a small proportion of landowners owning large amounts of land, while others own very little or no land. According to the latest available estimates obtained from the Agricultural Census 2010 (Pakistan Bureau of Statistics 2010), there were 8.26 million farms in the country. These farms were operating an area of 52.91 million acres. The distribution of farm area among small and large farms was highly skewed. Farms with fewer than 5 acres of land constituted 64 per cent (5.35 million) of the total private farms, but they farm merely 19 per cent (10.18 million acres) of the total farm area.[2] Conversely, farms of 25 acres and above in size, comprised only 4 per cent (0.30 million) of the total farms, but were spread over 35 per cent (18.12 million acres) of the total farm area across the country. It is important to note here that the landless poor (including sharecroppers and agricultural labourers) are not even included in such statistics aiming to capture landownership on the basis of farm sizes.

While there is evident variance within the farming community in terms of the amount of farmland they own. There are, however, different ways in which the Pakistani farming community may be categorized, which is highlighted in a more detail below.

Existing categories of Pakistani farmers

Farming communities are not homogenous within or across rural areas of Pakistan, and there are several ways in which different groups of farmers may be categorized.

The newly created state of Pakistan adopted two main land-tenure systems, with some regional variations, namely the *samindari* system and the peasant-proprietor land system (Qureshi and Qureshi 2004). Under the *samindari* system, *Jagirdars*[3] and *zamindars*[4] owned large tracts of land, which were cultivated with the help of landless tenants or sharecroppers. Many *jagirdars* and *zamindars* also provided their land in smaller plots to landless tenants working under hired intermediaries charged with supervising their work. These sharecropping tenants worked under different yield-sharing arrangements with landowners, depending on their capacity to share the costs of input. Tenants could, in turn, also be bifurcated into two distinct categories, i.e. occupancy tenants or tenants-at-will. While occupancy tenants had permanent, inheritable, and transferable rights to cultivate the *samindari* lands, tenants-at-will had no such legal rights. The other form of the land-tenure system that became evident in Pakistan was that of peasant-proprietorship. Peasants/farmers owned a more modest amount of land than *jagirdars* or *zamindars*, but they had landownership rights and cultivated this land mostly on their own, by using family labour, or hiring seasonal or daily wage labourers.

Other definitions can also be applied to distinguish between different categories of farmers. One such definition can be based on variances in economic status; rich farmers (those who rely extensively on the use of hired labour), mid-level farmers (who may use some hired labour, but mostly rely on family labour), and poor farmers (with insufficient land to even absorb family labour in its entirety, which in turn labours on the lands of others) (Mitrany 1952). Another distinction used to differentiate between varied categories of peasants is based on property relations. The status of a farmer can therefore be ascertained though the sort of claim that a particular farmer has to a piece of land, which may be that of permanent or long-term occupancy in the case of better-off farmers; whereas poorer peasants are often seen to be tenants-at-will who have short-term occupancy over land, mostly in the capacity of seasonal sharecroppers.

Subsequently, with the increasing commercialization of agriculture, leasing of land from both large and small landowners has also become increasingly common. Lease tenant farmers pay advance rents for the land they cultivate, bear all the costs of inputs, and hence retain the resulting crop yield in its entirety. The scale of lease tenant arrangements can vary widely, ranging from corporate farms having leased thousands of acres of land, to middle-sized farmers who lease land which requires them to hire additional labour for its cultivation, and those who can lease just enough land to absorb family labour. In practice, many poor farmers cannot afford to pay the upfront rents required to lease land, and it is usually households with alternative means of income,[5] who can afford to take land on lease.

There are also different types of agricultural labourers found in rural areas across Pakistan. Some of them seek employment in varying localities, where different crops are being sown or harvested, and are known as seasonal labourers, while others work on the land of larger landowners for remuneration in cash and/ or kind. While landless labourers do not have any direct access to land, they constitute, alongside poorer farmers, a major proportion of the rural population involved in agriculture. There are also agricultural labourers who work as permanent employees of large landowners, and are referred to as servitors since they are more like 'servants', except that they work in the fields instead of doing domestic chores.

There is frequent movement of poor farmers from one location to another. One form of such movement is mentioned above, where sharecroppers are shifted across different fields owned by larger landowners to prevent these poor farmers from claiming occupancy rights made available to tenants under existing legislation. Another form of movement also takes place in the form of seasonal labour migrations, where poor agricultural workers temporarily reside in other villages or districts to work on different crops on a seasonal basis. Agricultural labourers, sharecroppers, or even more affluent lease tenants, also move from one locality to another, where remunerations are relatively higher or more land or work opportunities are available (Gazdar 2005, 2007).

It is also relevant to recognize the particular circumstances surrounding farming communities due to the specific geographic locations in which they reside. While more location-specific characteristics are mentioned when discussing particular aspects of poor farmers lives in latter sections, it is instructive at this stage to provide a broader overview of developments of relevance to farming communities in Sindh and Punjab, the two provinces in which research was conducted for this book.

Farmers in Sindh

Sindh's economy is relatively industrialized, with agriculture contributing only 23 per cent of the provincial GDP. However, of the more than 33.46 million estimated people in the province, about 17.81 million were found to live in rural areas in 2006 (Pakistan Bureau of Statistics 2007). Later estimates have estimated Sindh's population to have reached the 42 million mark, half of which is still estimated to reside in rural areas (Government of Sindh 2012).

Moreover, while around 80 per cent of the rural population in the province was estimated to depend upon agriculture and its allied businesses, most of these people lacked access to their own land. While a median-sized[6] landowner in Sindh owned an estimated 28 acres of land, around 64 per cent of rural households owned no land and thus were compelled to work as sharecropping tenants or agricultural labourers for larger landowners (Hussein *et al.* 2004).

Sindh also has the most glaring disparities in terms of human development within Pakistan. A National Human Development Report prepared for

Pakistan, in collaboration with the UNDP in Pakistan in 2003, used human development indices (HDI)[7] to measure disparities within, and between, rural and urban areas of the country. This HDI ranking showed urban Sindh to have the highest rank, with an HDI of 0.659, which is higher than that for Pakistan as a whole (0.541). While urban Sindh ranked highest among the provinces due to its HDI achievements, rural Sindh had the lowest ranking in the country with an HDI of 0.456 (Hussain *et al.* 2003: 11), suggesting a much larger urban–rural disparity in Sindh, in comparison to the other provinces. The rural poverty situation has not changed significantly over more recent years either, since a more recent study also concluded that about two-thirds of rural households have been experiencing poverty for extended periods of time since advent of the new millennium (Pakistan Institute of Development Economics (PIDE) 2012).

The prevailing landownership situation in rural Sindh has also been shaped by irrigation works undertaken by the British during the nineteenth century. Colonial irrigation schemes significantly expanded pre-existing watercourses, bringing previously barren areas under cultivation (Cheesman 1996; Lieten and Breman 2002). The high demand for labour on the newly cultivable lands led to an influx of a large number of immigrants. Prior to these vast irrigation schemes Sindh was sparely populated but its population increased by up to 60 per cent between 1872 and 1911, from 2.2 million to 3.5 million people (*Gazetteer of Sindh* 1907, cited in Cheesman 1996). Yet, property rights in Sindh were not focused on provision of cultivable lands to poor landless farmers, but to *zamindars* (landowners), who were patronized to exert considerable political power over large areas (ibid.).

Traditionally, the *zamindars* used to pay revenue based on the crop shares handed over by the actual cultivators of the soil, known locally as *haris* (literally 'ploughmen'). *Haris* were basically tenants-at-will, many of whom had been cultivating the same piece of land for generations, while others were moved onto different fields belonging to the same landowner. These tenants generally paid rents-in-kind to landowners (*zamindars*), in the form of a share of their harvested yield. However, the crop share provided to landowners did vary significantly, ranging from one half to two thirds, depending on the quality of the land being tilled by the *haris*. Generally, agricultural inputs including seeds, dung used as fertilizer, the upkeep of animals required for ploughing, as well as human labour involved in cultivation, were the responsibility of the *hari*, while the *zamindars* provided land and water for the crops. In addition, *zamindars* also gave agricultural loans to *haris*, at different interest rates, for purchasing agri-inputs and meeting other household costs (Cheesman 1996).

Given the compulsion to secure productivity returns from heavy investments in irrigation schemes, and the fact that nearly two-thirds of the revenue being generated from the colonization of India by the middle of the nineteenth century consisted of rents from agricultural land, the British ruler tried to protect cultivators and motivate them to invest more in the agricultural production process (Stokes 1989). The Tenancy Act of 1885 in East Bengal was formulated with these imperatives in mind and introduced a series of tenant protection measures. This initial attempt was followed in the 1920s by a series of similar tenancy protection acts in other provinces under colonial rule. Such legislation was not provided in Sindh, however, and it was a few years after independence for such pro-tenant legislation to be passed within the province.

It was in 1950 that the Sindh Assembly passed the Sindh Tenancy Act. This legislative attempt aimed to articulate the duties of both tenants and large landowners, and also provided guidance for standardizing the division of crop yields between them. However, this legislative act was not adequately implemented in Sindh, due to manipulation by larger landowners, who continued to extract a much larger share of crop shares from their tenants.

The Sindh Tenancy Act of 1950 classified tenants according to the duration of their contracts in an attempt to introduce a basic level of tenure security. *Maurusi-haris* (permanent tenants) were provided the most protection, as their tenancy could not be revoked if they kept paying the agreed crop shares to the concerned landowners (Hussein *et al.* 2004). The *maurusi-haris*, however, needed to be farming on at least 4 acres of land which belonged to the same large landowner for a consecutive period of three years. To circumvent this legislative measure and prevent tenants from obtaining a legal right to cultivate a given piece of land, landowners began rotating tenants from one farm to another, after harvest had taken place, so as to prevent their tenants from becoming *maurusi-haris*. Furthermore, the provincial legislation tried to prevent *zamindars* from demanding free labour by the *hari* or his family. Yet, it lacked articulation of clearly defined tasks or limiting the number of working hours for tenants, allowing landowners to compel indebted *haris* and their families to work for them without remuneration (Ahmed 1984; Hussein *et al.* 2004). Thus, the broader objectives of the Sindh Tenancy Act to improve work relations between disempowered tenants and the large landowners whose lands they cultivated could not be achieved.

The introduction of 'Green Revolution' technologies, mechanization, and the subsequent market orientation of agriculture, did little to alter the circumstances of poor farmers, who own little or no land in Sindh. Despite considerable changes in the factors of production (high yielding varieties, irrigation, and mechanization) over the past few decades, large landowners are reportedly continuing to exploit landless tenants who work on their land (Hussain *et al.* 2003; National Coalition against Bonded Labour 2009; Zaidi 2001). Research done by the World Bank also indicates that the concentration of poverty in rural Sindh is highest among households where the household head is an unpaid family worker (60 per cent), *hari* (50 per cent), or owner-cultivator (40 per cent) owning fewer than 4.94 acres of land (World Bank 2002). In large landowner-dominated areas, where large landowners exert control over the local state apparatus, as well as the credit market, poor tenants/shareholders remain locked into a nexus of power and debt bondage with the larger landowners (Hussein *et al.* 2004).

Conversely, however, the World Bank contends that there is evidence that sharecroppers are better off than small-scale farmers cultivating the same amount of area, since the former have better access to credit and to input and output markets, gained due to their affiliation with large landowners (Mansuri and Jacoby 2006). Such an assertion helps justify the case for liberalizing land markets in Pakistan, to usher in corporate farming, and other similar measures, the impact of which on poor farmers, including landless sharecropping tenants, or agricultural labourers, is not without controversy (see Chapter 5 for more details).

Farmers in Punjab

Punjab is the most populated province in Pakistan. While the last population census was held in 1998, official estimates compiled in 2007 noted that the total population of the province was 86.45 million out of which 54.75 million people were still residing in rural areas. Subsequently, the Population Census Organization in Islamabad estimated the population of the province to have exceeded 95 million in December 2011, out of which over 30 million people were estimated to reside in urban areas, while almost 65 million people were still based in rural areas (Punjab Bureau of Statistics 2011: 299).

With the bulk of the population residing in rural areas, agriculture remains a vital sector with around 31 million acres under cultivation across the Punjab (Punjab Bureau of Statistics 2011: 47). As in the case of Sindh, it is also difficult to understand the current state of poor farmers in the province without the benefit of historical developments.

During the colonial period, western Punjab experienced extensive economic growth due to the vast process of canal construction and irrigation, where the British created the largest irrigation systems in the world. Development of nine major canal projects provided an increase in irrigated area of over 10 million acres. These canals were laid out in an area which was sparsely populated by semi-nomadic pastoralists. So the construction of perennial canals required extensive migrations from the more densely populated parts of the province. These 'canal colonies', as they later came to be known, were officially categorized as Crown 'Waste Lands', whereby the preceding claims of pastoral populations were not recognized, thus allowing the colonial state the power to determine how the land in this area was to be utilized (Ali 1987).

The colonial land grant schemes instituted in the Crown Waste Lands gave land grants to the existing landed rural elite in order to win their support for colonial rule, or else to agrarian classes that already held some land. A major proportion of land in each colony was allotted in the form of smallholdings, ranging in size from between 14 to 50 acres. The non-landed rural poor (agricultural labourers and servitors) were not meant to be provided land grants, even if they had also migrated to the canal colonies like the agricultural castes, and provided labour which was vital to ensure agricultural productivity within the newly cultivable lands. Due to colonial patronization, the upper- and middle-landowning peasant castes become compliant towards the British rulers, and they remained hesitant to join the nationalist cause until independence had become inevitable, and they were actively wooed by nationalist leaders to support the creation of Pakistan (Ali 1988).

It is important to note that colonial land grants initially gave agricultural castes occupancy rights only. The colonial state had several requirements which it expected peasant land grant recipients to fulfil, such as maintenance of field boundaries and water channels. Land grantees kept demanding proprietary rights, and their increasing agitation led the British government to concede by passing a new Colonization of Government Lands Act in 1912, which gave them propriety rights. This concession implied that colonial state's retreat from the originally envisioned interventionist role of actively encouraging agricultural development. This new Act provided peasant land grantees proprietary rights after they had spent a qualifying period as an occupancy tenant. However, revoking the prior requirement of residence of the grantee in colony villages to supervise the agricultural production process enabled the phenomenon of absentee landlordism. Instead of residing on the land granted to them, and supervising the cultivation process, colonial land grantees began renting out their lands to intermediaries, and living off of these rents in urban areas (Ali 1987, 1988).

Therefore, despite introducing irrigation technology into the Punjab, the British were not able to effect meaningful change in the agricultural production processes or the underlying social organization hierarchies of production in the canal colonies. Instead their selective provision of land grants to peasant castes excluded many other categories of poor farmers from availing themselves of these schemes. Dilution of the original schemes also enabled land grantees to become less engaged with the agricultural production process and instead rely on rents extracted by exploitative intermediaries. Whatever surplus reached the government in the form of land revenues was not spent on improving the lives of poor farmers either, but instead utilized to finance military expenditures, and meet other expenses of the colonial administrative structure. The primary beneficiaries of these colonial land grant schemes were either peasant grantees, or their intermediaries who actually controlled the land and oversaw the work of sharecroppers and agricultural labourers (Ali 1987). In effect, these short-sighted policies encouraged greater concentration in landownership, and increased levels of disparity in the agrarian social structure.

Based on anthropological work in the Punjab just prior to the 'Green Revolution' in the 1960s (Eglar 2010), five distinct categories of cultivators could be found in rural Punjab. In the first category were large landowners who had more land than they could cultivate themselves, and needed others (poorer and landless farmers) to help cultivate it. In the second category, were landowners who owned sufficient land to work on themselves, and rented out their spare land. In the third category, were landowners who cultivated their own land but did not rent out land, even if it they had more land than they could self-cultivate. In the fourth category, were landowners with around 7 acres, the agricultural yield of which was not sufficient to support them,[8] and so they rented land from other landowners and cultivated it alongside their own. In the fifth category were those who were landless or owned very little land, described as tenants-at-will, and these were sharecropping on land belonging to larger landowners. These categories were, however, disrupted by the 'Green Revolution'. Many tenants-at-will became displaced, either leaving agriculture or becoming agricultural labourers, as landowners took back land given out under sharecropping arrangements and began cultivating it with the help of waged agricultural labourers and subsidized machinery such as tractors (Herring 1983).

Yet, the dominance of elite landowners from the colonial period continues to this day. An analysis based on a unique combination of household surveys, archival data, and family genealogies, in the district of Sargodha, empirically demonstrates that political dominance established 150 years ago persists due to the nature of the developments which have taken place in the post-colonial state (Cheema *et al.* 2009). One argument put forward to explain this phenomenon is the time lag, of around two decades after independence, before the political nexus between the state and the landed rural elite could be challenged. While the principle of universal adult franchise was recognized by the Constitution drafted in 1956, it was not until 1970 that a general election was held in the country. Therefore, it was only in 1970 that the preferential rights granted to landowners by the colonial administration could be influenced in practice by the electoral space, which then became available to all adult citizens, including villagers. Even then, the absence of effective land reforms, the imposition of another two decade of martial law (in the 1980s and 2000s), and increasing dependence on pro-growth policies favoured by donor, which rely on larger landowners to boost agricultural productivity, have helped preserve the dominance of the landed elite.

One evident example of the glaring disparities, and concentration of power and assets in the hands of the few, is the unequal distribution of land across the Punjab. This inequality is particularly severe in the southern[9] and western[10] districts of the province. In 2000, 5 per cent of owners in the South held approximately 37 per cent of owned area, and 7.8 per cent of owners held 46 per cent of owned area in the west (DFID 2009). Moreover, there is also a decrease in sharecropping in favour of leasing out land to tenants in the southern and western districts of the province, where the sharecropped area as a percentage of farm area has declined by 15 per cent, from over 30 per cent in 1980 to 15 per cent in 2000 (ibid.: 84). The major decline in sharecropped land is accompanied by increasing fixed-rent leasing patterns as well as land being resumed by landowners for self-cultivation through agricultural labourers and increased reliance on mechanization.

It is important to note that land leasing does not offer significant opportunities to the landless poor farmers, who remain unable to pay upfront rents, and bear all the costs of production themselves. Sharecroppers losing access to land are considered prone to becoming increasingly dependent on daily waged on-farm and off-farm work, which is not well remunerated, and this trend increases their probability of becoming poorer (DFID 2009; Cheema *et al.* 2008). Given the persisting level of inequality in landownership, increasing land fragmentation, accompanied by the decline in sharecropping, the accessibility of cultivable land by the landless poor does not seem to be improving.

Realities and challenges facing poor farmers

Poverty in rural areas is closely correlated to landownership. A longitudinal survey comparing changes in overall rural poverty levels between 2001 and 2010 notes that rural poverty decreased from 31.3 per cent in 2001 to 27 per cent in 2010 (PIDE 2012). A regional analysis, however, shows that in southern Punjab and Sindh there was a net increase in poverty between the 2001 to 2010 period. There was, in particular, a higher incidence of chronic poverty common among households headed by less educated persons, and having no ownership of land and livestock.

The scourge of food insecurity also affects those who have either little or no land.[11] Toor (2010), for instance, has argued that better access to land helps achieve better nutritional outcomes. On the other hand, Altaf (2010) highlights the precarious conditions of poorer farmers, who make up a great majority of the country's cultivators, who are described to be using more than two-thirds of their income on food, some in Sindh spending as much as 87 per cent. This expenditure on food leaves the rural poor with very little income to spend on health expenditures, or on the education of their children. The need to meet unavoidable costs pertaining to illness, or death and marriage, often push poor households into debt. The fact that these poor rural households are unable to educate their children leads to the perpetuation of their inter-generational dependence on larger landowners.

Moreover, for three consecutive years (2010, 2011, and 2012), floods have caused unprecedented levels of damage across vast portions of the country-side. The 2010 monsoon floods have been assessed to be the worst since 1929. According to the National Disaster Management Authority (NDMA) the rains/floods have affected over 20 million people in the country (over one-tenth of Pakistan's population). Approximately 1.6 million homes were destroyed, along with many thousands of acres of crops and agricultural lands across all four provinces (Asian Development Bank and World Bank 2010). In 2011, floods again hit the country, impacting around 5.4 million people, with the most severe impacts on Baluchistan and Sindh (United Nations Office for the Coordination of Humanitarian Affairs (UNOCHA) 2011). In 2012, floods hit Sindh again, where 2.8 million were affected in comparison to only 890,000 people affected in Punjab, and the damage was less severe in other provinces (NDMA 2012, 2013). These floods did not directly hit districts where I con-ducted research for this book, however, the indirect impact of these recurrent flood disasters was felt across the country in the form of population displacement from flood-hit districts. There was subsequent movement of the population from areas such as the Muzaffargarh district in southern Punjab, coming to settle all the way in the central Punjab district of Gujrat, bringing in both seasonal workers and agricultural labourers. While the research questions posed within this book were not designed to ascertain the impact of the flood on poor farmers in particular, research findings did invariably reveal state institutions responding to the needs of larger landowners instead of poor farmers, as will be discussed in more detail in the next chapter.

Challenges facing women involved in agriculture are also daunting. In most developing countries, large-scale surveys and agricultural censuses collect property-related information only by household, without disaggregation by gender (Agarwal 1995). The case in Pakistan is no different, where small-scale surveys and research studies have to be used to assess the situation concerning women's access to land. Findings that have emerged from such sources indicate that few women own arable land, and even fewer effectively control it, across the rural areas of the country (Brohi 2010; MHHDC 2003; Mumtaz and Mumtaz 2012).

There are further problems with the collection of reliable gender-disaggregated data, due to which women's real contribution to the national economy is not adequately reflected in official statistics. The Federal Bureau of Statistics data on women's employment, including its Labour Force Surveys, also do not capture the actual extent of women's work contribution. However, the Mid-Term Development Framework 2005–15, prepared by the Planning Commission of Pakistan, notes that over 70 per cent of rural women engaged in work relating to agriculture and livestock (Government of Pakistan 2005). Thus, besides the burden of raising children and looking after their households, women also share the economic burden of ensuring their household's survival. Women look after livestock and help their husbands in the fields, but also work as casual seasonal labour. Until the 1950s, women's labour was confined to specific jobs in agriculture, such as planting, harvesting, and threshing rice; harvesting wheat; and picking cotton. At present, however, in rural areas of Sindh and Punjab in particular, women are engaged in almost all agricultural operations. They are, however, severely underpaid and being forced into accepting low paid labour in the form of daily wage labourers. There is legislation, and state land redistribution schemes, aiming to help women's access to landownership, but these too have lacunas that are identified in subsequent chapters.

Conclusions

Overall, poverty in rural Sindh and Punjab remains inextricably linked to prevailing land tenure arrangements, which have been shaped through colonial and post-colonial regimes. While large landowners wield significant political and social power, the majority of the rural population has a dearth of opportunities to improve their lives. A significant proportion of the rural populace can be included in the category of poor farmers, including not only those with very small parcels of land, but also landless sharecroppers, seasonal labourers, daily wage agricultural labourers, and women involved in agriculture. In large landowner-dominated areas, where large landowners exert control over the local state apparatus as well as the credit market, the poor farmers often get locked into a nexus of domination and debt bondage with large landowners. As sharecropping opportunities are diminishing in favour of leasing across the rural landscape, poorer farmers who cannot afford to pay upfront rents either become landless agricultural workers, or they abandon agriculture completely and struggle to find livelihoods in the off-farm sector or in the unsustainably growing urban areas of the country. A series of natural disasters has made life even more difficult, and food insecurity is now a particular problem even for poor landless families who are directly involved in the cultivation process. Given the variance of challenges confronting poor farmers, an attempt is made in the following three chapters to examine how state and donor policies for agricultural development and rural poverty alleviation aim to address these multidimensional problems, and also how some poor farmers are themselves reacting to the circumstances in which they find themselves in contemporary Pakistan.

Notes

1 The East India Company ruled over the subcontinent from 1757 until the Indian rebellion of 1857, when the British Crown exerted direct control over the Indian territories.

2 The number of poor farmers is also growing over time according to the Agricultural Census Report, since farmers with fewer than 5 acres of land went up from 58 per cent in 2000 to 64 per cent in 2010 (Pakistan Bureau of Statistics, 2010).

3 *Jagirdars* were large landowners who were given revenue-free land estates by the government.

4 *Zamindars* were large landowners who had to pay a land tax to the government.

5 Such alternative means of income can include money sent back to villages by family members working in cities for instance.

6 Median is a type of average used in statistical analysis, which arranges a given set of values in order and then selects the one in the middle.

7 Supported by the UNDP and informed by its HDI which ranks countries according to their development accomplishments, the NHDR also used HDI to undertake provincial and district-based development rankings within Pakistan. The NHDR HDI ranking included three main components, each affecting, in one way or another, a human being's life by way of his/her access to 'means' and/or desired 'ends' which were: (1) health, (2) education and (3) income.

8 This was before the advent of mechanization, and the use of high yield varieties of seed and increased use of fertilizers and pesticides, ushered in under the 'Green Revolution' from 1958 onwards.

9 Southern Punjab consists of seven out of 38 districts of the province, namely: Rahimyar Khan, Bahawalpur, Bahawalnagar, Multan, Lodhran, Khanewal, and Vehari.

10 Western Punjab consists of seven out of 38 districts in the province, namely: districts of Mianwali, Bhakkar, Khushab, Layyah, Muzzaffargarh, D.G. Khan, and Rajanpur.
11 Pakistan's landless poor account for an estimated 60 per cent to 75 per cent of the country's rural populace (Anwar *et al.* 2004; Toor 2010).

4 Ascertaining impacts of state institution and policies on poor farmers

This chapter considers the role of a range of state institutions, and varied policies espoused by them, to specifically ascertain their implications for poor farmers. The chapter begins by assessing how poor farmers' concerns are reflected in political manifestos of major political parties, and then draws attention to recent legislative attempts which have relevance for them. In this regard, recent legislative attempts to encourage corporate farming and protect plant breeders' rights are examined with particular reference to the potential adverse impacts of such legislation. Conversely, failed attempts to bring land reforms back on the policy agenda through formulation of a draft land redistribution bill, and challenges confronting recently approved legislation aiming to address gender disparities in landownership, are also discussed. Implementation of a state land distribution scheme for the landless poor in rural Sindh is considered next. This chapter then turns to assessing specific institutional policies, including those concerning land revenue management, provision of agricultural extension services, and offering of agricultural subsidies, to determine the extent to which these measures are beneficial for poor farmers. The chapter will also illustrate how different state officials exhibit biases against already marginalized poor and landless farmers.

This chapter also highlights how some of the above-mentioned legislation, as well as specific state policies, such as the implementation of user charges for veterinary services or revenue generation attempts within the irrigation and land revenue departments, are subjected to donor influence. A further attempt is also made to demonstrate how the state chooses to operationalize donor advice and the corresponding effects of resultant policies and programmes for different categories of farmers. On the other hand, the state's reluctance to concede to particular donor demands, such as curbing subsidization in the agriculture sector, is also discussed to reveal contradictions between the stated rationale used to justify subsidization, and who existing subsidization schemes actually end up benefiting.

The Pakistani state and poor farmers

Poor farmers in Pakistan remain susceptible to a range of vulnerabilities due to the lack of effective measures to protect their interests. This is despite the fact that Article 38(a) of the Constitution of Pakistan specifies the need for the promotion of the social and economic well-being of the people (Government of Pakistan 1973: 7). It clearly states:

> The State shall secure the well-being of the people, irrespective of sex, caste, creed or race, by raising their standard of living, by preventing the concentration of wealth and means of production and distribution in the hands of a few to the detriment of general interest and by ensuring equitable adjustment of rights between employers and employees, and large landowners and tenants.

These constitutional provisions however have not been translated into reality. Even where legislation exists, it is not effectively implemented. Consider, for instance, provincial tenancy laws, such as the Sindh Tenancy Act of 1950, or the Punjab Protection and Restoration of Tenancy Rights Act of 1950, which govern the legal relationship between the large landowners who own, and the tenant farmers who occupy, rural land. Tenancy legislation divides farmers into two categories: 'occupancy tenants', who have a statutory right to occupy the land, and 'simple tenants, or tenants-at-will' who occupy it on the basis of a contract with their landlord. While tenants-at-will can be easily evicted from land, occupancy tenants do have more rights under tenancy laws. Yet, in practice, large landowners can easily dodge these tenancy laws since they exercise great influence over the revenue and other local officials. When a large landowner takes on a tenant as a sharecropper, this arrangement is often not entered into the official revenue record, despite the fact that these documents are meant to keep a record of not only who owns the land, but also who is cultivating it. Therefore, landowners can deny with impunity all tenancy rights and protections that are legally the tenant's due under the available tenancy legislation.

The altering of land records persists even in areas where large landholdings have declined, and traditional landowners have lessened their control over those who cultivate their lands. In Gujrat, for instance, where landholding sizes are fairly modest and even poorer farmers are better off in comparison to other districts in southern Punjab, I found inaccuracies in cultivation records. This discrepancy also appeared in the land records for Azaz Ali (not his real name), a prominent politician and legal expert in the country.

Azaz Ali is a senior politician affiliated with one of the most prominent political parties in the country, and he has served as a senator as well as in other senior government positions. He is also a prominent lawyer and a constitutional expert.

Kotpur (not real name) is Azaz's ancestral village, where his family has lived for generations. His grandfather was a *Zaildar*, a native official appointed in charge of a unit of colonial rural administration of Punjab in British India. Azaz himself resides in Lahore, which is where he had been contesting his electoral seat from, instead of doing so from his ancestral village. His lack of dependence on local voters was identified as the primary reason for his aloofness by the local residents of the village, many of whom pointed out that all the state facilities in their area, such as electrification, and the local primary schools for girls and for boys, were provided during the tenure of governments, formed by rival political parties. Being such a successful lawyer, Azaz is not dependent on his agricultural income either, so he simply leases the land he owns in the village to get upfront revenue without having to supervise sharecropping arrangements.

Azaz has leased out his land to Arif, whose family had been in a sharecropping arrangement with Azaz's elders for generations. This lease tenant, Arif, did not reveal the exact amount he was paying for the 17 acres of land he had leased from Azaz, but mentioned that the terms were favourable.

Arif (not real name), the lease tenant, is himself fairly well off, and he also made no attempt to conceal this fact during my interactions with him. During one of my visits to the village, for instance, after I had finished doing some interviews at his *dera*,[1] a car and driver came to pick me up. Arif noticed me wave to the driver asking him to bring the vehicle closer to the *dera*, and commented that he had become a bit confused thinking that I was issuing instructions to his driver, since he too owned a Toyota Corolla.[2] Arif's affluence was not confined to such self-affirmations alone. Arif's status was also reaffirmed by Haji *Sahib*,[3] himself a prominent landowner of the village, who observed that strong landowners (Azaz) have strong leaseholders (Arif) as well.

As mentioned above, Azaz himself is considered a prominent person not because of his landholdings, but due to his other accomplishments as a lawyer and politician. Moreover, the fact that Arif was cultivating a sizeable amount of land on lease implied that he was not a poor or disempowered tenant farmer. Despite Arif's personal affluence in the village, however, it was surprising to see no mention of Arif as the cultivator on the land revenue records maintained by the village *patwari*. Instead these official land records stated that Azaz was cultivating the land himself.

In this specific case, it did not seem that Azaz's declaration of being a self-cultivator was meant to exploit Arif. Instead it seemed more like a safety measure to ensure that Arif would not try to make any claims on Azaz's ancestral lands, especially since Azaz did not come to the village too often. However, this sort of distortion of cultivation records appears to be common, and would perceivably make it very difficult for lease operators or sharecroppers, especially those with less affluence, to have recourse to legal justice in the case of a dispute with landowners.

The situation for poor sharecroppers, who cannot afford to pay upfront rents to lease land, and meet all the costs of production on their own, is even more tenuous. Sharing the costs of inputs with landowners, and often relying on them for the sale of their crops, gives rise to more opportunities for dispute. However, the imbalance of power between landowners and sharecroppers, and the sharecroppers' lack of legal status on land cultivation records, can invariably result in decisions favourable to those who own the land, rather than the poor cultivators.

In the worst cases, sharecroppers can become so heavily indebted to landowners that their movement is restricted and their entire family is compelled to work to pay off the debt (Arif 2008; Zaidi 2001). A Bonded Labour System Abolition Act was put in place in 1992, which declared debt bondage to be illegal. In 2002, the Sindh government also made some provisions regarding a tenant's debt to the landowner, such as restricting large landowners arbitrarily deducting payments from the sharecropping tenant's portion of produce, and also forbidding the use of unpaid labour from family members of the sharecropping tenant[4] (Arif 2008). Yet these efforts have not been able to prevent bonded labour in districts such as Umerkot, which was one of the districts where my fieldwork was conducted.

The year preceding my visit to Umerkot, a district court judge had freed 78 bonded labourers, including 31 children and 21 women, from an agricultural farm where a landowner had been detaining them due to loan amounts totalling Rs 900,000 (Dawn 2010). This was just one of several reported cases of multitudes of bonded labourers who still need to be freed by judicial intervention in Sindh on a regular basis. Many more cases of bonded labour in agriculture do not even make it to the courts.

There is as yet no effective legislation available to offer adequate protection to seasonal or daily wage labourers in the agricultural sector, including women. A minimum monthly wage requirement of Rs 7,000 was stipulated for 'unskilled' workers in the Labour Policy of 2010 and subsequently raised to Rs 10,000 in both Sindh and Punjab in 2013 (Government of Pakistan 2010; Government of Punjab 2013; Government of Sindh 2013). However, the government does not enforce this measure in the 'informal' agricultural sector, since agricultural production is not considered a part of the formal economy and is therefore not subjected to labour inspections. A majority of the landless labourers I interviewed in the districts of both Sindh and Punjab were receiving well below this amount (this issue will be taken up in detail while discussing salaries of daily wage earners at a corporate farm in Chapter 5).

I also found several agricultural servitors who were not receiving conventional salaries at the end of each month, but would instead be paid only the amount needed for some unavoidable expenditure, while the rest of their money would be used to make deductions from advance payments they had received at the time of commencing work for a landowner.

In the village in Gujrat, I interviewed over a dozen such servitors. While these servitors cannot be described as being locked into conditions of debt bondage, since they have freedom of movement and their entire families were not being coerced to work to pay off their debts, the conditions in which some of them resided were appalling. I met a 26-year-old man, employed by a local landowner in Gujrat, who was living with his pregnant wife and two young children in a windowless bare brick room within a livestock pen inhabited by a half-dozen livestock. His pay for working from dawn to dusk was Rs 4,500 per month, and 40 kilograms of grain, and he was still trying to pay off a Rs 24,000 loan he owned to his employer. In fact, all the other servitors I met had salaries ranging from Rs 3,000 to Rs 6,000 per month.

The lack of formal legislation recognizing the rights of agricultural labourers and servitors provides them no recourse to justice if there is a dispute over wages. There are also no formal contracts concerning agreed work expectations, or documentation of financial transactions between landowners and their employees. The lack of literacy and the power differentials between landowners and landless labourers would further complicate the actual implementation of any potential contractual obligations.

NGOs have sporadically tried to provide representation to poorer farmers by undertaking advocacy for them, but this effort has not yet resulted in any significant results. Despite repeated calls for the unionization of agricultural labourers, which could provide them collective bargaining power, no union for this vast segment of the rural labour force exists. Moreover, since agriculture is considered part of the informal rather than the formal sector, its labour force remains exempt from inspections by labour department officials. There is also not much evidence of other specific measures meant to protect poor farmers, including landless labourers or women, as subsequent sections of this chapter demonstrate.

a. Manifestos, legislation, and poor farmers

It is insightful to consider how the political manifestos of major political parties aim to contend with the challenges of agricultural development and rural poverty alleviation, and to also assess recent legislative attempts, which are of direct relevance to poor farmers.

Landowners have exerted a major influence in politics since the time of independence in 1947. However, the emergence of new political forces, such as the urban-based Muttahida Quami Movement (MQM) and religious parties, is considered to be decreasing the strength of large landowners in the legislature. For instance, it is pointed out that in Punjab, the country's most populous province, the number of national lawmakers from landlord families shrank to 25 per cent in 2008 from 42 per cent in 1970 (Tavernise 2010). However, the presence of large landowners is still formidable enough to prevent the coalescence of any pro-land reform block within the parliament. The first Prime Minister elected by General Musharraf's government in 2002, Mir Zafarullah Jamali, soon after becoming Prime Minister, categorically claimed that there was no further need for land reform in the country, and that land redistribution had become a dead issue within Pakistan (Ahmed 2003). The Prime Minister under the Pakistan People's Party (PPP) government, which completed its tenure in 2013, also belonged to a prominent landowning family from southern Punjab.

Moreover, the two largest political parties have evidently not given redistributive land reforms any serious reconsideration in electoral manifestoes since the past few decades. The PPP, the only democratically elected party to have tried implementing land reforms in the 1970s, seems to have lost any further aspirations to appropriate land from larger landowners and distribute it among poor farmers (Aftab 2011). The PPP's manifesto for 2008, on the basis of which the political party assumed the reins of government for a five-year term (until 2013), did not even acknowledge the problem of landless poor farmers (PPP 2008). Instead, this manifesto has a section on accelerating agriculture and rural growth, which places emphasis on 'improving productivity and crop diversification, agricultural markets and exports'. Acknowledging that the 'Pakistani peasant, mired in poverty and debt has to be rescued from the morass of despair', the PPP manifesto in 2008 had placed faith instead in 'a bold policy which ensures that the private sector provides key inputs and services – such as credit, fertilizer, pesticides, extension, marketing, seeds, tractors – in a timely manner and at competitive prices' (ibid.: 9). Such market-based solutions to addressing rural poverty are quite a step away from the 1970 PPP manifesto which had declared 'breaking up of the large estates to destroy the power of the feudal landowners is a national necessity' (ibid.: 6).

The PML-N electoral manifesto, which brought the party to power for the third time in May 2013,[5] rhetorically admits that 'marginal adjustments in development policies will not address the issue of mass poverty'. It thus calls for a paradigm shift to 'evolve pro-poor growth strategies that will change institutions and local power structures in favour of the poor, by giving them greater access to productive assets such as land' (PML-N 2013: 32). It also recognizes that small farmers as the 'real back-bone of the rural economy', who need 'access to knowledge, inputs and markets' (ibid.: 29). Yet no other explicit mechanisms to achieve these aspirations are mentioned except the need to revitalize corporate agriculture to overcome limitations faced by small landowners, and to set up land development corporations with majority equity of the poor and managed by professional managers. Not much information was provided about these envisioned land development corporations, nor had such corporations been formed until the writing of this book. It was however clear that the PML-N too had chosen to ignore the need for land redistribution and increasingly rely on market-led approaches to agricultural development instead.

The last military regime in Pakistan passed the Corporate Farming Ordinance (CFO) in 2001, allowing foreign firms to lease land in Pakistan for 99 years, extendable for another 49 years. A minimum requirement for land leasing was set at 1,500 acres, while no maximum ceiling was placed on leaseholdings. The CFO also allowed remittance of 100 per cent capital investments, as well as profits made from corporate farming undertaken on leased Pakistani state land 2001 (Board of Investment 2004). The CFO 2001 itself had also purported lofty aims, such as seeking to improve agricultural productivity and profitability through the use of the latest production technologies and adequate expertise, particularly for exports (ibid.). However, besides aiming to encourage local investors, the CFO 2001 has been criticized for facilitating a 'global land grab', whereby richer countries can lease or purchase land from poor countries, and utilize its other scare resources, most importantly water, to grow food and other needed agricultural products (Kugelman and Hathaway 2010).

In Pakistan, countries interested in corporate farming have been mostly Gulf countries,[6] including the United Arab Emirates. However, the information on state agricultural land lease deals is fragmented and often contradictory. In May 2009, the Ministry of Investment offered 1 million hectares (2,471,054 acres) of state-owned land for long-term investment to the Emirates Investment Group, for using the increasingly stressed land and water resources of Pakistan to grow crops meant for export, at the risk of diminishing the availability of these required natural resources for poor Pakistani farmers (Toor 2010). There are also reports of ongoing negotiations, mostly with other Middle Eastern investors (including Qatar, Bahrain, and Saudi Arabia), to lease state land in the Thar and Cholistan regions of Sindh and Punjab, and in Baluchistan (Kerr and Bokhari 2008; Sadeque 2009; Settle 2013). Yet it remains unclear if foreign agricultural investors leasing Pakistani state land will be required to provide food for consumption within Pakistan, or if they will offer suitable employment opportunities to the local rural populace at their high-tech farms. While there was no evidence of UAE or other Middle Eastern-based corporate farms operating in districts where my research was conducted, corporate farming by affluent local farmers has already commenced, the implications of which are explored in the following chapter.

Pakistan is obligated to provide protection for the development and marketing of seeds, due to being a signatory to the World Trade Organization's negotiated Trade-Related Aspects of Intellectual Property Rights (TRIPs) Agreement.[7] This has been done by the formulation of the Plant Breeders' Rights (PBR) Bill 2012 (IPO-Pakistan 2012). The Bill categorically endorses the need to give effect to sub-paragraph (b) of paragraph 3 of article 27 in part II of the TRIPS agreement (WTO 1995: 12), which states:

> Members shall provide for the protection of plant varieties either by patents or by an effective *sui generis*[8] system or by any combination thereof.

While emphasizing the need for granting patent rights, the PBR Bill 2012 does not acknowledge the United Nations Convention of Biological Diversity 1992 (UN 1992: 1), which categorically stresses the need for protecting farmer and community rights to seeds.

Without providing specific means to ensure the protection of indigenous genetic assets being used by disadvantaged and marginalized small farmers, the PBR Bill 2012, in its present form, could instead be used to begin restricting farmers from continuing the centuries old traditional systems of the storage, sharing, and multiplying of indigenous seeds. Many farmers in Sindh and Punjab complained that their forefathers were able to grow numerous crops, whereas they have increasingly been compelled to practise mono cropping by larger landowners, with whom they have sharecropping arrangements. Since sharecroppers often take loans from landowners, they are compelled to pay heed to landowner discretions concerning the crops which should be grown, and the type of agricultural inputs they require. In fact, the very need for using expensive agricultural inputs like fertilizers and pesticides has been impelled by the increasing trends towards mono cropping of cash crops, which is an underlying reason for their indebtedness in the first place. Being compelled to purchase branded and/or genetically modified seeds will imply a further increased cost of input, adding to the loan burden of poor farmers, including sharecroppers.

The experiences of other developing countries like Kenya support this assertion, where PBR rights seem to have been predominantly applied by the foreign-owned commercial exporters of flowers and vegetables to underpin commercialization and boost agricultural exports, rather than address the concerns of Kenya's poor farmers or improve the quality and lower the price of seeds for the crops they grow and rely on for food security (CIPR 2002).

Representatives of local and international civil society organizations,[9] working on land rights and environmental issues in Pakistan, argue that the main beneficiaries of the existing PBR Bill 2012 would be landowners with the capacity to engage in commercial farming and the multinational corporations making inroads into the seed industry in Pakistan.

The draft form of the PBR Bill 2012 is to be submitted to the parliament so its final approval remained pending until the end of 2013. However, the government's Intellectual Property Organization (IPO) seems poised to get this legislation passed in the near future. There was no indication of the IPO taking steps to address the above-mentioned concerns of relevance to poor farmers.

Besides the above-mentioned broader legislative measures, the most directly relevant legislative attempts of significance to poor farmers are land reforms efforts, and one such attempt to formulate a new land reform bill was undertaken in the recent past, to which attention is drawn next.

b. A failed attempt to bring land redistribution back on the legislative agenda

While earlier land reform attempts in 1959, 1972, and 1977 are criticized for not having been implemented effectively, the latest land reform proposal drafted in 2010 fizzled out even before it could achieve the status of implementable legislation. The Redistributive Land Reforms Bill 2010 was drafted by the urban-based political party, the Muttahida Qaumi Movement (MQM).[10] While the latest draft bill aimed to address some of the loopholes which had diluted the effectiveness of earlier land reform attempts, the political party that tried to introduce this bill lacked the legitimacy and support needed within the national assembly to have its proposal passed into law.

Analysis of the MQM bill itself indicates it has set out rather stringent measures in comparison to the earlier land reform bills. It proposes lowering the landholding ceiling much further. While the 1959 reforms had declared a land ceiling of 500 acres for irrigated or 1,000 acres for un-irrigated land, which was brought down to 150 acres for irrigated and 300 acres for un-irrigated land in 1972, the MQM bill proposes a ceiling of 36 acres for irrigated or 54 acres for un-irrigated land (MQM 2010). The bill further stipulates that this land-ownership ceiling must be placed on land occupied by a family instead of individuals, meant to prevent evasions experienced during previous attempts, when individuals transferred land to the names of their family members to prevent its confiscation.

The MQM's draft bill has aimed to provide compensation to the owners of the land at rates determined by a land reform commission (ibid.). It, however, has not explained how resources for remunerating existing landowners are to be secured. Donors, like the World Bank, could potentially be approached for providing the required funding to help pay off landowners whose land is to be redistributed. The World Bank is, in fact, known to have lent money for this purpose in other countries, including Mexico, Colombia, Brazil, the Philippines, Thailand, and Indonesia (Pereira 2005). However, the experience of implementing such market-assisted land reforms is not positive, since the loans provided by donor agencies are transferred by the state onto poor people, who have to repay expensive loans, often from harvests from poor soils, given that landowners often choose to sell them marginal and ecologically fragile plots (Borras and Franco 2010; Pereira 2005).

Moreover, the very motivation of the MQM-drafted bill has been challenged by not only the establishment but also several analysts as a mere political ploy by a primarily urban-based political party to make inroads into rural areas. For example, on the one hand, a senior provincial official in the Legal and Parliamentary Affairs Department in Lahore dismissed the MQM bill as being hypocritical, claiming that the MQM needs to first reconsider its involvement in land mafia deals and encroachments in urban areas like Hyderabad and Karachi, which are its immediate constituency, before it takes up the agenda of achieving more equitable land distribution in rural areas. Other stakeholders including civil society representatives, and politicians (including the former finance minister under the Zulfiqar Bhutto government in the 1970s, when the 1972 and 1977 land reform bills were introduced), also dismissed the MQM bill as a political tactic aiming to do no more than create a rural constituency in areas dominated by other mainstream political parties with support from local landowners.

The MQM thus remained unable to secure support from other prominent political parties in the parliament for this proposed bill. Conversely, religious political parties have vocally opposed this bill referring to the 1990 Supreme Court ruling that declared land reforms 'unIslamic'[11] (Pakistan Press International 2010). The proposed MQM bill tried to counter the religious elements in its draft bill by asserting that there is no provision in Islam for absentee landlordism,[12] and that large landowners demanding a share of produce from a tenant, in return for the use of land owned without physical involvement in the cultivation, is *haram* (forbidden) (MQM 2010). The proposed bill also claims that Islam enjoins the equitable distribution of wealth and economic powers, and abhors their concentration in a few hands. Such assertions, however, did not convince the Jamiat Ulema-e-Pakistan (JUP), which spearheaded the opposition to the MQM bill and prevented it from becoming an act of law (Gazdar 2011).

It is thus not only politicians and legislation, but also other institutions, including religious elements within the judiciary, which help preserve the status quo of landownership. The above-mentioned Supreme Court decision has, however, been challenged through a judicial petition, as is discussed in the following sub-section.

c. The judiciary's role in preserving uneven landownership patterns

Despite delivering a blow to the land reforms when the Supreme Court declared placing any ceiling on landholding 'unIslamic' in 1990, the judiciary has shown some sporadic signs of redressing the plight of poor farmers. One such instance was when the judges of the Supreme Court of Pakistan, in 2003, issued a court ruling in support of a peasant whose land in Punjab had been seized by a military brigadier[13] (Siddiqa 2006). Although it was a military officer who had brought the case to the court against the poor farmer, the court instead ruled in favour of the poor farmer, and also cited excerpts from John Steinbeck's *Grapes of Wrath* in its verdict (Steinbeck 1939: 161), including the following:

> And the great owners, who must lose their land in an upheaval, the great owners with access to history, with eyes to read history and to know the great fact: when property accumulates in too few hands, it is taken away.

Such cases of judicial activism, however, remain few and far between, and poor farmers often lack the resources or the know-how to bring their cases to the court system.

Nonetheless, some recent public interest litigation has been filed within the apex courts, which explicitly draws attention to the distorted rural political economy in the country and the underlying structural causes of the marginalization and deprivation of the rural masses. This litigation attempt is based on two petitions, the electoral reforms and the land reforms petitions, filed in 2012 by seven petitioners. These petitioners include two politicians (the General Secretary of the Workers' Party Pakistan, and the Senior Vice President of the National Party, which are both registered leftist political parties);[14] two labour union officials (General Secretary of the Confederation of Trade Unions of Pakistan and the President of Pakistan Trade Union Federation); the President of the Pakistan Kisaan (Farmers') Committee; the President of the Democratic Women's Association, and an Associate Professor in the Faculty of Law and Policy at the Lahore University of Management Sciences (Minto and Minto 2012a, 2012b).

Both these petitions referred to the prevailing culture of feudalism in the country. The first petition asked the Supreme Court to address this by revising its earlier objections to land reforms. The second petition sought to move the Supreme Court to compel the executive to initiate electoral reforms to counter the political influence of landowners.

The first petition was filed under Article 184 (3) of the Constitution of Pakistan, 1972, pertaining to fundamental rights of citizens, and has challenged the Supreme Court's earlier decision (in 1990) which had upheld an appeal against a Federal Shariat Court ruling that had initially declared land reforms instituted by the democratic government of Zulfiqar Ali Bhutto to be 'unIslamic'. The pending petition filed in the Supreme Court is based on two arguments. First, it asserted that neither the Federal Shariat Court, nor the Supreme Court, have the jurisdiction to alter fundamental provisions of the Constitution of 1973, which itself had sanctioned land reforms. Second, the petitioners pointed out that the 1990 judgment maintaining land reforms to be unIslamic, needs to be reconsidered since Maulana Taqi Usmani, the chairman of the judicial bench at the time, had hesitated in calling its decision a final one, due to the lack of sufficient scholarly input representing all schools of Islamic thought on the subject (Minto and Minto 2012b). Maulana Usmani had, himself, also conceded that protection of private property applied only if it was obtained, in the first place, through legitimate means. The petitioners have thus argued that the earlier judgment did not take into account the fact that in most cases, the landownership of big landowners was illegitimate to begin with, since they came about as largesse extended to them by the British colonists for services rendered; services that often comprised of conspiring with the British against the local populace (Sadeque 2012). The petitioners concluded their arguments by pointed out that the matter of land reforms being declared 'unIslamic' may have had more validity in a 'truly Islamic state', which was able to fulfil all the specified duties pertaining to the welfare for all its subjects, which was clearly not the case in a country like Pakistan with its immense disparities, and thus there was no justifiable reason to prevent the redistribution of wealth through redistributive land reforms (Minto and Minto 2012b: 22). The petition also made the point that the existing feudal landowning system in Pakistan makes the existing electoral process largely meaningless, which deprives people of their constitutional right to freely choose their political representatives. It is this particular issue that the second petition takes up more specifically.

The second petition was also filed under Article 184 (3) of the Constitution of Pakistan, 1973, and it asked the Supreme Court to enable more effective participation of citizens in the political affairs and governance of the country (Minto and Minto 2012a). The argument made in this petition was that existing practices and processes employed in holding elections remain unconstitutional, since they have had made it very difficult for anyone except a small minority of people to engage in the political process. Clause 8 of the petition asserted that under existing conditions, 'the overwhelming majority of the citizens of this country can never aspire to achieve their legitimate social, economic and political objectives by participating in governance unless the unconstitutional practices currently underlying the process of elections are restrained or remedied' (ibid.: 5). The inferences being made here pertained to the expensive culture of campaigning, electioneering, and contesting elections, which has remained unchecked and over time evolved into making the process of contesting elections exclusionary and heavily tilted in favour of retrogressive forces (such as the landed rural elite) which seek to perpetuate the status quo which guarantees their hegemony over the politics of the country. Besides offering suggestions to regulate election activities, to curb spending during elections, and to make election procedures more transparent, the petition specifically highlighted the political and electioneering culture in areas dominated by feudal landholdings, to be highly subject to manipulation whereby: 'the voter is not always entirely aware of his/her rights and duties regarding the process of voting itself' (ibid.: 31). In turn, the petition stressed the need for voter education to prevent manipulation of voters by landowners to preserve their power and their vested interests.

It is worth noting how both the petitions conceptually reinforced each other especially with reference to the power of large landowners which is considered to be distorting not only economic but also the political opportunities available to citizens within the country. While the hearings concerning the land reforms petition remained pending until early 2014, the Supreme Court held hearing in April 2012 concerning the petition challenging the election system, and the court has since issued a formal decision as well (Supreme Court of Pakistan 2012). The concerned Supreme Court judgment issued several instructions to the Election Commission of Pakistan to make election processes more transparent and effective. However, the extent to which these recommendations had an impact on the election process during the general elections in 2013 remains open to dispute. Several political parties disputed the election results, accusing the incumbent government of election rigging as well as other irregularities (Haider 2013). What is more relevant in the context of this book however is the judgment's implications for land reforms, given that the election reforms petition was prepared alongside the land reform petition, based on the rationale that unless land reforms are implemented, fair elections cannot take place within rural areas in particular.

The Supreme Court judgment on the petition challenging the election system did not really take up the land reforms issue in a significant manner. It only mentioned the distortions caused to the electoral process due to the political influence of large landowners, or 'feudals' in three instances. First, the judgment acknowledged the demand of contending political parties advocating need for appointing federal rather than provincial presiding officers at polling stations, since they are less susceptible to feudal pressure, but this issue was not squarely addressed in the judgment itself, and its resolution was instead left to the Election Commission. Second, the problem of feudals influencing voters to back them, or their candidates, was addressed by the need for a targeted voter education programme, also to be undertaken by the

Election Commission. The need for computerizing voting was dismissed, using the argument that computer balloting may enable feudals to amass computer devices and use them to cast votes themselves to the complete exclusion of the voters, which was the third reference to feudal interference within the electoral process. The need for effective land reforms to be implemented within the country for fairer, or more democratic, elections to take place in Pakistan was however not addressed in the Supreme Court's decision.

One of the reasons why the judiciary remains reluctant to reinstate the possibility of land reforms, is that this would put the institution at loggerheads with mainstream political parties, many of whom are also prominent landowners. The potential for litigation on its own to change the situation on the ground is limited. There are more chances of the above discussed attempts being subverted rather than being endorsed and effectively used to influence social and political discourses in favour of poor farmers.

d. Addressing the problem of women's lack of ownership of land

While redistributive land reforms have not received serious political attention or support, the PPP-led government did secure passage of the Prevention of Anti-Women Practices Bill, 2011 through the National Assembly, during its recent tenure. Besides providing a minimum legal benchmark to penalize those who engage in customary practices which force women and girls into marriage to settle personal, family, or tribal disputes,[15] this law has criminalized the act of forcibly denying a woman her right to inheritance, including land (Khan 2011).

Assessing the potential for legislation aiming to protect women's right to land, it is necessary to keep in mind the fact that women in rural areas more often voluntarily forfeit their right to landownership due to social pressures. Thus criminalization, only pertaining to the forcible confiscation of their land, is dependent on their coming forward to engage in criminal proceedings in the first place.

Although lack of landownership by women is not confined to rural areas alone, it is particularly lack of control over agricultural land that is being discussed here. Given the fact that ownership of land empowers people, the prevailing lack of effective control over land assets is directly correlated to the continuing lack of Pakistani women's empowerment. Without addressing the underlying imperatives due to which women themselves forfeit their right to inheritance, especially cultivable land, the Anti-Women Practices Bill 2011 faces a range of potential hurdles to achieving its goal of improving women's ownership of land.

Women's economic contribution in Pakistan generally remains unacknowledged, both inside and outside the home, and within the agricultural sector as well. Rural women are actively denied control over assets, particularly cultivable land[16] (Brohi 2010). In 2001, the Pakistan Rural Household Survey found that women owned only 2.8 per cent of agricultural plots across the country (cited in World Bank 2005). This is despite the fact that the Constitution and laws of the country allow women to own and inherit property. Article 23 of the Constitution of Pakistan (Government of Pakistan 1973) makes the following provision:

> Every citizen shall have the right to acquire, hold, and dispose of property in any part of Pakistan.

Moreover, religion also grants inheritance rights to women. According to legal interpretations by the Sunni sect of Islam, applicable to over 80 per cent of the country's population, a wife (or wives) gets one-quarter of the share if there is no child, otherwise they get one-eighth. A mother gets one-third share if there is no child, otherwise one-sixth. A daughter gets half the share of a son. In the absence of a son, the daughter gets half the share of the inheritance and if there is more than one daughter, they collectively get a two-thirds share. *Shia* laws,[17] applicable to approximately 15 per cent of the population, also prescribe similar proportions, with a few exceptions, such as daughters without brothers get higher shares of the inheritance (Brohi 2010).

In reality, however, property rights in Pakistan are governed by an amalgam of civil law influenced by English common law principles,[18] Islamic law as interpreted by jurists, and customary laws. While civil laws dealing with the ownership and transfer of property are gender neutral, inheritance rights are subjected to Muslim personal law.[19] Pakistani civil courts have also enforced *Sharia* law for inheritance purposes, but they do not generally ensure if their decisions are being implemented. The government machinery including the land revenue department also seems reluctant to address the sociocultural barriers preventing implementation of women's land rights (SDPI 2008). The complexity of legal requirements and bureaucratic processes, alongside the disempowerment of women within Pakistani society, further limit the potential of women stepping forward to claim and subsequently to acquire whatever rights are legally granted to them.

The land reform legislation adopted in the country so far seems to have a built-in gender bias, because they do not particularly emphasize the need for redistributing land to women. However, land reforms by Field Marshal (R.) Ayub Khan in the late 1950s, for example, did provide some inadvertent advantages for women by limiting holdings for individuals, and compelling many large landholding individuals to transfer property in the name of women within their families as well, in order to keep their landholdings secure (Brohi 2010). Even legal ownership of land did not allow women to exert actual control over land, or to derive economic benefits from it.

I also found evidence of women across villages in Sindh and Punjab still being denied access to land, as well as control over land, despite possessing its legal ownership. At the time of the death of a family head, the property is supposed to be transferred to all the children, and this property transfer is meant to be accordingly reflected in the land revenue records. While land is supposed to be inherited by daughters as well, the brothers often compel sisters to legally transfer their land over to them before their sisters are wed and have to leave the house. Even if women legally own land, they are not given control over cultivation decisions, nor are the earnings derived from the land, which is in their name, handed over to them. Most of the villagers I spoke to across the districts where research was conducted confirmed that women only owned land on paper, and that sisters in particular would hand over the land inherited by them to their brothers. The villagers justified this practice as a social custom and argued that brothers help provide dowry for their sisters,[20] and they look after their mothers after the father passes away. It was, therefore, their right to be rewarded by being able to retain control over their family land. Some landowners pointed out that they had given gifts or compensation to sisters of the same value as their land. However, these same landowners admitted that the men in the family wanted to retain direct control over agricultural land. Trying to justify the prevailing practices of denying women their rightful share of inheritance, a landowner in the Kotpur village pointed that while girls forfeit their right to land when they leave their parents/guardians, on the other hand they join families where their husbands have taken the land of their own sisters, and therefore the equation balances out.

Villagers were not the only ones who defended this prevalent practice. An agriculture department official in Gujrat opined that there was a misconception regarding women being coerced into relinquishing their landownership rights. He said that women genuinely gift land to their brothers without any element of coercion, since their brothers not only bear the costs of arranging dowries for getting their sisters married, but also provide a social safety net for unforeseen crises, including estrangement with, or even the death of, their husbands.

A senior police official in Gujrat admitted that landownership of women was rare in practice. A district-based agricultural official, however, drew class distinctions between women's ownership of land. He said that women tend to fight over land in larger families where high economic costs are involved. Conversely, in poorer families, where land sizes are small and women are more dependent on their brothers for social protection, male members of their families claimed these women were 'happy' to give their land in lieu of their continued protection in case of some unforeseen calamity such as the death of their husbands.

While a handful of cases where women owned land in approximately a half-dozen villages I kept revisiting during my fieldwork were found, women's ownership of land was not necessarily class based. For example, I found women in large landholding families to have acquired land, such as in the case of the farm manager at a large corporate farm in southern Punjab, who mentioned he had given land to his sister. Another example was found in Azaz's family, the prominent barrister and landowners in Kotpur village in Gujrat, whose family had given their only sister her share of inherited land, who in turn had chosen to sell her sizable property, the proceeds of which she also retained.

Conversely, Haji *Sahib* another prominent landowner in Kotpur village in Gujrat, was opposed to giving property to women, and mentioned that his sisters had not been given land, and he did not intend to give his daughters land either. Several other landowners also felt that it was against traditional values for a sister to ask for land belonging to her family to be sold, given to her, or given to her husband. Those who demanded this right were threatened with social exclusion. One landowner narrated an instance of a local woman from their village who had taken a dowry from her brothers at the time of marriage, but then demanded, and thereafter sold, her share of inherited land, which in turn led local villagers to sever ties with her to such an extent that when she returned after a number of years for a younger brother's wedding, she got a very insipid welcome not only by men but also by the other women of her family.[21]

On the other hand, I also saw poorer women owning land, and also exerting some element of control over it. I interviewed a poor widow in the southern Punjab who had inherited land from her husband, and claimed she had complete freedom over whom she would give this property to before her passing. A *Masalahat* Committee[22] Chairman in Gujrat also confirmed his involvement in cases where women had contested and gained their right to sell their share of land, and some had even 'sold their land for pennies' just to deprive their brothers who were opposing them. Others had given land to some brothers, while denying it to others with whom they did not get along. However, he admitted that men were also involved in this process, since some brothers were backing their sisters' decisions.

In many other instances, however, where women seemed to own land according to official records, it was either their husbands or other household male members who were cultivating the land, while also controlling the benefits obtained from it. In the Kotpur village in Gujrat, for instance, Allah Ditta (not real name) did not own any land but his wife had inherited 8 acres, since she was the only child of her parents, and now Allah Ditta was cultivating this land to run their household. It was Allah Ditta who was investing money to cultivate the land, and it was he who controlled the earnings derived from this land, not his wife, even though the land still officially belonged to her. Yousaf (not real name), a 45-year-old farmer in the same village, was cultivating 4.5 acres of a 9-acre land parcel he had inherited when his father passed away. Yousaf's brother was cultivating the other half of the land. While according to ownership records, their mother and sister also had a share of the property, the brothers were effectively managing the land, and neither of the women received revenues generated from the land that was still legally in their names.

The village *patwari*, as well as several of his colleagues in the Land Revenue Department, further confirmed that women in the area did not exercise their ownership rights, even if the property legally remained in their names. Thus while according to the land records women owned nearly 500 acres out of approximately 2,200 acres in the Kotpur village (in Gujrat district), this owner-ship did not lead to significant empowerment. Several villagers also conceded that while several women owned property on paper, they generally did not get a direct share of the remuneration from their property. In some cases, married sisters were getting a share of the harvested yield, but this is considered a sign of their brother's generosity rather than an intrinsic right based on landownership.

Patwaris agreed that fraud does take place to prevent women from owning land. I was informed that in cases where a woman refuses to forfeit her right to land, landowners have been known to bring in a woman posing to be their sister, with forged documents, declaring that she had forfeited her right to the inherited land. However, most women who would come to forfeit their right to inherited land were said to be doing so willingly. Some of them came and said on record that they wanted to gift the land to their brothers, while others claimed to have been compensated for giving away their land, even if they had not received direct payment for the land, given that this was legally a more secure means for brothers to consolidate their hold over family land. A few of the *patwaris* pointed out that land revenue officials did ask women if they were relinquishing their right out of free will, and if the women confirmed this to be the case then officials had no legal right to prevent the land transfers.

While there are multiple reasons why women are denied their land rights, all of them are related to the lack of empowerment of women within the patriarchal system prevalent in the country. Earlier research has pointed to the lack of information about legal, economic, or political rights, the violence that is embedded in customary practices which deny women land rights, and restricted mobility, as some of the collective factors which deny women opportunities of owning and accessing assets and resources (RDI 2009; SDPI 2008). Another significant barrier to women for laying claim to their land rights is their lack of access to information concerning land registration and transaction procedures, the formal process for taking possession of land, and/or retaining their share. A majority of rural women are also unaware of judicial recourse that is available to them if their land rights are denied.

Passage of a new bill (Prevention of Anti-Women Practices Bill 2011) in the National Assembly, after two failed attempts, does mark a positive step towards providing women the clear-cut option of legal protection to safeguard their property rights. However, women's rights advocates are concerned that the bill is narrowly focused on punitive aspects, such as prescribing imprisonment and fines in cases where a court has determined that women were being defrauded of their right to property. The bill does not take into account, or address the reasons why many women are still unable to report offences against them in the first place.

Concerns about the new bill include the lack of adequate mechanisms to ensure more effective reporting of inheritance related crimes, and their expeditious hearing within courts of law (Ali 2010a; Brohi 2010). During my research, none of the police officials and *patwaris* I interviewed seemed aware of the Prevention of Anti-Women Practices Bill 2011, despite the fact that they are the frontline officials when it comes to dealing with problems concerning inheritance issues. A senior official in the Law and Parliamentary Affairs Department in Lahore pointed out that Pakistan had female officers in the police, and in other departments, and suggested that the same be done in the land revenue department to help overcome existing gender-based distortions in landownership. However, the above-mentioned legislation makes no such provision.

The above discussion has illustrated the prevailing attitude of the state in addressing the plight of poor farmers, ranging from a discussion of relevant legislative attempts to an analysis of official manifestos of political parties, and the lacklustre role played by the judiciary in this regard. Due to the prevailing rural political economy in Pakistan, most state institutions offer little practical opportunities for the empowerment of poor farmers, including women involved in agriculture. The need for redistributive land reforms also remains on the backburner, despite the isolated attempt by the MQM to reintroduce legislative debate concerning land reforms, an attempt which was impelled by opportunism resulting in a failure to rally needed political backing. Other legislative attempts to empower poor farmers, including women, exhibit both lack of political and suffer from design flaws which undermine their effectiveness. Given the inability of the state to put in place a broader enabling legislative environment to ensure more equitable land access, and the disinterest of political representatives to address this issue, more specific interventions aiming to provide state land to poor farmers will now be considered.

Distribution of state land

British colonial administrators, and subsequent governments in Pakistan, have repeatedly allocated state land to a range of beneficiaries to achieve different goals, ranging from providing patronage to secure loyalty, trying to boost economic productivity and even to alleviate rural poverty (Alavi 1973; Ali 1987; Talbot 2007). The state also grants government land under resettlement schemes to people affected by dam construction or other development projects.

Instead of resuming ownership of land from large landowners, the ruling class can formulate rather contentious policies aiming to gain the favour of different constituencies with the distribution of state land, strengthening the politics of patronage. A state land distribution policy announced by the Chief Minister of Punjab, for instance, created a heated debate in 2011 when he announced plans to hand over 25,000 acres of district managed forest areas, and surplus land revenue department lands, to create 25-acre plots for each young graduate in agriculture to help boost agricultural productivity (Hasnain 2010). Environmental concerns regarding deforestation by taking over forested areas administered by the forestry department caused a delay in the implementation of the scheme. One positive development of the delay was the refinement of the earlier ad hoc announcement into the formulation of an

'Agro-Forestry scheme', which proposed a model aiming to plant trees, inter-cropped with vegetable and forage crops, on the government provided land. Whether this provincial scheme will help boost agricultural productivity by the intended beneficiaries, without further depleting the scant forest cover within the province, remains to be seen. Yet since it is rare for poor farmers to achieve graduate degrees, the scheme did not seek to directly benefit them. Nor had this scheme made any special provisions concerning land allocations for women.

A range of policy and implementation failures explain the muted effect of state land redistribution policies on alleviating land inequality. Primarily, state land redistribution policies appear not to have effectively targeted the eradication of landlessness in the first place. The fact that a large segment of the rural populace remains landless indicates that state land redistribution has not been able to adequately address the widespread landlessness in rural areas.[23] Nonetheless, it is instructive to see exactly how these policies are implemented on the ground, and to identify specific reasons for them being unable to achieve their objectives. The ongoing implementation of the Benazir Landless *Hari* Scheme during field visits to Sindh in 2011 provided me an ideal opportunity to undertake such analysis.

a. Benazir Landless **Hari** Scheme

Soon after assuming power across the country in 2008, the PPP-dominated provincial government in Sindh announced it was going to grant over 212,864 acres of government owned agricultural land, half of which is prime irrigated land, and the rest rain dependent, to landless farmers across 17 districts of the province. Although the amount of state land being distributed could do little given the scope of landlessness in the province,[24] it did claim to have adopted an innovative approach to earlier state land distribution schemes. The Benazir Landless *Hari* Scheme gave overriding priority to women since 70 per cent of the intended 5,800 beneficiaries of this scheme were meant to be women (Government of Sindh 2008a, 2008b).

Moreover, the provincial government's land distribution scheme acknowledged that merely distributing land is not enough, since it is essential to also provide additional support to poor farmers in order to ensure that they become successful cultivators. Thus, the Sindh government planned for the beneficiaries of state land would be fully supported through a support package for a period of at least two years, until such time as they attain sustainable livelihoods (Government of Sindh 2008b). For this purpose, the government decided to collaborate with three major NGOs in Sindh, namely Sindh Rural Support Program (SRSP), the National Rural Support Program (NRSP), and Thardeep Rural Support Program (TRDP), which undertake community development schemes across the province, with a major focus on provision of micro-finance.

These NGOs are more aptly described as quasi non-governmental organizations (QUANGOs), which were established with government support and receive funding from a range of government ministries such as health, agriculture, livestock, and the environment ministries (Ghaus-Pasha and Iqbal 2003). Moreover, all three of these Rural Support Programmes (RSPs) also receive substantial funding from donor agencies like the World Bank, which is one reason why they rely on market-based prescriptions for poverty alleviation, such as reliance on micro-finance (RDI 2009).

Within the context of the Benazir Landless *Hari* Scheme, the three mentioned RSPs (NRSP, SRSP, and TRDP) took on the task of developing support packages for scheme beneficiaries to ensure they could effectively develop the granted land. The support package devised for this purpose included micro-loans for provision of agricultural inputs (seeds, fertilizers, etc.), and creation of linkages between beneficiaries and relevant government entities such as the land revenue, irrigation, and agricultural extension departments. In addition, beneficiary households were also provided micro health insurance, providing hospitalization and accident coverage, family nutrition kits, poultry, and fruit plant saplings, in order to further help improve their living standard (Hasnain 2010).

One of the background papers prepared by the government of Sindh for the Benazir Landless *Hari* Scheme pointed out how earlier land grant schemes had not attempted to target the landless in a transparent manner, and hence land grants were made, more or less, either at the discretion of the revenue staff, or on the basis of political patronage (Government of Sindh 2008b). The Benazir Landless *Hari* Scheme claimed it had not only developed an effective institutional mechanism at the provincial and district level to ensure transparency and prevent nepotism and corruption, but that it also planned to use an effective database[25] to enable the provincial government to target genuine *haris* (Government of Sindh 2008a).

Yet there has been criticism of this scheme by poor women in the areas where this scheme was implemented, ranging from claims that the land being allocated is not suitable for cultivation, to complaints of patronage.[26] The national press has also quoted some of these complaints. Consider, for instance, a statement cited in a prominent English daily newspaper saying:

> In most of the cases, it [land] was given to sisters and close relatives of feudals in the area and the small percentage that was given to landless women farmers was either not suitable for farming or was located far away from the residence of that farmer
>
> (*The News* 2011)

Other NGOs tasked to oversee this distribution scheme pointed out that, contrary to the guidelines issued at the provincial level, civil society (the aforementioned RSPs) were not very involved in the verification of the land identified by revenue department officials at the district level.

(PDI 2009)

To probe how land unsuitable for cultivation had been earmarked for distribution under the scheme, or why undeserving beneficiaries were allocated land, I met with some of the officials, implementing agencies, and intended beneficiaries of this scheme in the Umerkot and Badin districts. Senior management within one of the RSPs, which was facilitating the government of Sindh on this project in Umerkot, further informed me that while thousands of people applied for the scheme within the district, only a minor proportion of beneficiaries were selected, which predictably indicated that the amount of state land available was much less than the number of eligible candidates for the scheme. However, what was more revealing about the way this scheme was implemented were admissions by several RSP personnel that many of the scheme beneficiaries who got land were those who had political patronage. The bureaucracy, and parliamentarians, were said to be interfering in the selection process. Many politicians had apparently suggested names of beneficiaries to revenue department officials. Land revenue officials, in turn, did not bother with undertaking comprehensive poverty assessments of given communities, and simply issued land grant orders to many beneficiaries who were either sharecroppers for, or relatives of, influential landowners in the district.

Several beneficiary households in Umerkot also criticized the criteria applied by local government officials to select beneficiaries. For instance, one criterion was that of proximity to ensure that intended beneficiaries could practically cultivate the land, while the other criterion was that of need, determined by the fact that they do not own any land. The criterion of proximity was however used selectively so that those living in closest proximity to state land were given precedence over other village residents who lived further away in the village, even if they were more deserving. A senior RSP manager in the district estimated that out of 132 beneficiaries identified in the first phase of allocations, only 24 were completely genuine and had absolutely no land. He further mentioned that during the ongoing second phase of allocations, more beneficiaries (304) have been selected in their area of operations, and even less genuine cases are coming forth. These estimates are not to be taken lightly, given that the concerned RSP has obtained independent detailed data from government selected beneficiaries, using its own criteria of their sources of income, amount of livestock, number of children, schooling levels, access to land, etc. The RSP officer who had been facilitating my discussion subsequently with a group of 14 beneficiary families later informed me that only three of them were found to be completely landless.

Out of 304 beneficiaries in the second round of land grants by the scheme, the RSP was giving cheques for aid to only 92 of them. Genuine need or poverty was, however, not the criterion for RSP's decision to work with the 92 scheme beneficiaries. Rather it was the fact that the others either had not obtained the complete set of required forms from land revenue officials, ownership of the land allocated to them was disputed, or they had been given land which was not suitable for cultivation. The RSP criteria for support was more concerned with ensuring the success of their programmatic intervention, so they selected only those beneficiaries who had complete paper work and cultivatable land, rather than being concerned about whether the beneficiaries were genuinely landless poor farmers who could be assisted with completing the paperwork.

The RSP criteria of success for the project also did not require increasing female empowerment by ensuring that men would consult them in making decisions about cultivation, or to ensure that women retained control over the income generated from the land allocated to them. Group discussions with beneficiary households in Umerkot also confirmed that men of the household were still taking the responsibility of cultivating the land, and making all major decisions concerning the cultivation, even if the land was allocated to women in their households.

The RSP, which had been tasked to help the government of Sindh implement the Benazir Landless *Hari* Scheme in Badin district, was also most focused on selecting beneficiaries who had managed to secure land titles from the government, and received parcels of land which were suitable for cultivation. The Badin based RSP management also noted that farming households with some political connections, and even pre-existing parcels of land, were being provided complete land titles for the most cultivable land under the Benazir Landless *Hari* Scheme. Conversely, some of the deserving beneficiaries of the scheme, who had not yet managed to obtain all the required paperwork for the land allocated to them, were being excluded from the RSP programme of support.

Neither of the RSPs in Umerkot or in Badin was seen to be undertaking intermediation between these poor farmers and the state machinery to help ensure that the poorer and more deserving beneficiaries of the Benazir Landless *Hari* Scheme were also being provided complete land registration documents, and being allocated cultivable land, so they could actually be helped to become sustainable farmers. Instead, both RSP interventions focused on working only with beneficiaries whose land grant paperwork was in order and who had been given land that was suitable for cultivation, even if many of these beneficiaries already possessed land, but had managed through patronage to become beneficiaries of a government scheme aiming to target landless households.

Therefore, despite the categorical goal of the Benazir Landless *Hari* Scheme to specifically target the landless poor, it is in fact allowing undeserving beneficiaries to gain access to state land. Involving the RSPs in implementation did not seem to make the scheme more effective, since such organizations also became preoccupied with meeting their own project objectives of ensuring the success of scheme beneficiaries they had begun supporting, rather than ascertaining whether these selected beneficiaries were actually poor landless farmers or not.

Thus, the Benazir Landless *Hari* Land Scheme also encountered effective targeting and political interference problems despite NGO/QUANGO involvement, which are typical problems, which hinder the implementation of many government interventions. Yet, this unfolding experience does not seem to have deterred the PML-N in placing similar faith in state-led land distribution schemes. The PML-N election manifesto on the basis of which the party won the 2013 elections also endorsed the need for provision of state lands to landless farmers and women. It is important to note that the PML-N's self-described 'land reform programme' states its intention to 'reclaim and irrigate additional land' for allotment to poor farmers (PML-N 2013: 31), rather than tackle the concentration of landownership in the hands of the rural elites.

Government departments, their policies for agricultural development, and the resulting impact on poor farmers

Prior research has highlighted a range of institutional biases towards favouring large landowners, often to the disadvantage of those with access to little or no land (Gazdar 2007; Hussain 2005, 2008). Significant attention has been drawn, for instance, towards the implementation of 'Green revolution' policies, which offered subsidized agricultural support to larger famers instead of poorer and landless farmers in the 1970s (Alavi 1973; Niazi 2004). Irrigational policies have also been criticized for directly correlating water rights with landownership. The amount of land someone owns determines how much water he or she gets. Landless people, including farmers, and women who do not own agricultural lands, thus have no voice in how water is distributed (Hussain 2008; Hussain *et al.* 2003). Ethnographic research also illustrates how large landowners are backed by other key government officials such as land revenue (*patwaris*) or police officers to the disadvantage of poorer farmers, in cases of sharecropping disputes[27] (Hussain 2005; Malik 2008).

State institutions are, however, not static entities, and the policies espoused by them are subject to change. Comprehensively ascertaining the impact of government policies involving the vast array of institutions operating within the rural sector in Pakistan at present is beyond the scope of this research. However, the following subsection focuses on describing some salient and ongoing interventions of relevance to agricultural development, and ascertaining their combined implications for poor and landless farmers.

a. Irrigational problems confronting poor farmers

Pakistan is described as a highly water-stressed country, which is already using 97 per cent of its surface water resources, and mining its groundwater at unsustainable rates, to support one of the lowest agricultural productivities in the world, per unit of water, and per unit of land (Kamal 2009).

As mentioned earlier, colonial interest in boosting cash cropping helped Pakistan acquire the largest irrigation system in Asia, but it also led to numerous challenges such as water logging and salinity,[28] and the pressure of maintaining an elaborate irrigational infrastructure to match the growing need for irrigational water. Water rates, known as *abiana*,[29] varied from crop to crop and from canal to canal, which increased from time to time to meet the expenditures incurred on maintenance of the irrigation system. However, with the passage of time, although irrigational expenditures kept increasing, the government began subsidizing water rates aiming to create incentives for increased agricultural production. The government has justified its subsidization of irrigational water as being a 'pro-poor' measure. Yet, it has done little to lessen the inequity of water sharing between larger and smaller farmers. Water rights in Pakistan remain tied to ownership of land, whereby those without access to sufficient land remain unable to draw enough water to irrigate their fields (Hassan 2008; Shafique 2008).

Despite substantial budgetary input, the irrigation system in Pakistan has been suffering from worsening operational problems and incapacity to manage the available water resources efficiently. One response to the shortage of canal water has been investment in private tube-wells, and the rise of groundwater markets. While tube-wells have provided a reliable source of irrigation, groundwater levels have been declining over the years, and diesel-powered tube-wells are also expensive to run, making them a very costly alternative for poor farmers (Jacoby *et al.* 2004). When poor farmers, including sharecroppers, have to pay for the cost of diesel to power pumps needed for using groundwater to cultivate their crops, this increases the burden of cultivation expenses, and compels many of them to become more indebted to large landowners and intermediaries such as commission agents.

Under donor pressure, especially that of the World Bank, to curb public expenditure and lessen inefficiency, the federal government established Provincial Irrigation and Drainage Authorities in each province to look after the operational systems of irrigation networks by the late 1990s. The revised model envisaged forming Farmer Organizations (FO), comprising around a dozen members, to be elected by farmers irrigating land from a given watercourse. In turn, these FOs were meant to supervise the supply of water to farmers, become responsible for operation and maintenance, and collect levies and water charges (Hassan 2008). The stated objectives of these irrigational reforms, including FO formation, were to reduce the corruption of rent seeking public officials, introduce transparency and efficiency, and to bring more landowners into the revenue collection net by making evasion more difficult.

The irrigation department officials I interviewed, however, pointed to several problems in FO formation. They lacked confidence in FOs' ability to become democratic entities, as they are still being dominated by larger landowners. It is larger landowners who were considered to be exerting control over who should be selected to be part of the FO, or be allowed to engage in its decision-making processes. FO members were also reportedly hesitant to register complaints against other farmers within their communities who were not making *abiana* payments, or were engaging in illegal breaches, since they had to rely on these same people for support in order to remain FO members. A Punjab Drainage and Management Authority (PIDA) project official in Lahore, directly involved with the process of FO formation, also pointed out that women only vote for FO members, and do not really become part of FOs themselves. This is another problem with FO formation since such a mechanism can offer little prospects to enhance women's role in agriculture.

PIDAs have modified the procedure of *abiana* collection, and announced application of flat rates, to simplify the process of assessment and to facilitate more effective revenue collection. In Punjab, this flat rate amounts to Rs 85 per cropped acre during the *Kharif* season and Rs 50 per acre during the *Rabi* season.[30] These flat rates were put into effect in 2003 (Hassan 2008; Shafique 2008). This was done despite the fact that the low level of flat rate charges applied uniformly to all socio-economic groups of farmers, instead of differentiated irrigational charges, implying that the incentives for the users to improve water use efficiency are further reduced. Flat tax rates are much easier to manage and implement. However, poor farmers are said to be worse off since a nominal service charge remains insufficient to make water management more efficient, and whatever water is available continues to be utilized by large landowners, despite the formation of entities such as FOs.

Another PIDA official in the Punjab, however, claimed that the *abiana* collection record of FOs is impressive, since around 60 per cent to 80 per cent of FO members are now paying water charges. Yet, he admitted that other aspects of FO performance are more difficult to assess, especially whether or not they had prevented water theft, or lessened water disputes between farmers sharing the watercourses. PDMA tried to collect benchmark data in 2005, but the irrigation department could not provide them consolidated information, so it is hard to tell if FOs formation, and the use of privatization of irrigational management, had resulted in its stated goals of not only improved efficiency, but also preventing inequitable water use. Also, no disaggregated data could be found to indicate whether water availability to poorer farmers had increased or diminished since the implementation of these reforms. The manipulation of poor and landless farmers within the FOs is understandable given the underlying rural political economy in Pakistan, and unless the very nature of tenure relations and landownership patterns is altered, it would be difficult to ensure that entities like FOs can actually become participatory and democratic decision-making entities.

Besides the challenge of managing irrigational resources in an equitable and efficient manner, provincial irrigation department officials were implicated in a serious charge of mismanagement and corruption while responding to the massive 2010 floods, which affected nearly 20 million people across the country. In the aftermath of the flood, numerous reports began emerging from around the country, including southern Punjab and Sindh, of politicians and large landowners having pressured irrigation authorities to breach dykes in a manner that would save their own lands, at the cost of drowning out the lands and villages of poor farmers (Ahmed 2011; Walsh 2010). Widespread public anger had since compelled the Supreme Court to issue directives to the provincial irrigation departments to investigate and present details of personnel involved in such breaches, and departmental inquiries to this effect were ongoing at the time of this research.

The above incident points to the continued problem of collusion between large landowners, who are often politicians or have influence with them, and state institutions such as the irrigation department, which enables them to tamper with public infrastructure, even in the times of disaster, and at the cost of the lives and livelihoods of a multitude of disempowered people.

b. Taxing agriculture and its effect on poor farmers

While revenue collection within the irrigational sector has the less ambitious goal of cost recovery for operations and maintenance, the potential for revenue generation from agricultural income is much greater. Before focusing on the agricultural sector itself it is, however, useful to take a broader view concerning the problem of generating needed revenues for the state through an effective taxation system.

Pakistan has one of the lowest tax-to-GDP ratios in the world, which has varied between 8.5 per cent to just over 9 per cent in recent years (Cheema 2012). This situation has evident implications on the ability of the indebted state to effectively fulfil its mandated responsibilities and functions. The problem of tax evasion starts at the very top of the ruling elite. More than 60 per cent of Pakistan's cabinet and two-thirds of its federal lawmakers paid no tax during 2011 (Associated Press of Pakistan 2012; Cheema 2012).

Besides outright tax evasion, taxation in Pakistan is not progressive, so the poor have to share as much of the brunt of revenue generation as the rich. Due to the rampant tax evasion, the government relies on indirect taxation such as sales taxes, which do not distinguish between poor or rich citizens. Sales tax has in fact become the major contributor in the federal tax receipts with a 48.3 per cent share in total collection followed by direct taxes (33.4 per cent), customs duties (12.8 per cent) and federal excise duties (5.5 per cent) in the first quarter of financial year 2012–13 (FBR 2012: 3).

The case for agricultural income tax, specifically, follows from the low contributions of agriculture to the government's tax revenue (Chaudry 1999). Agriculturalists themselves have, however, been resisting a direct income tax claiming that the sector is already overburdened with indirect and implicit taxes in the form of low agricultural commodity prices. Donor agencies like the World Bank have been pressuring the government to replace the land revenue system with agricultural income tax for increasing revenue generation from this sector. The World Bank has been advocating the need for increasing the level of taxes raised from agricultural income by pointing out how agriculture is a large part of Pakistan's economy, and by not taxing agricultural income effectively, the tax burden on other sectors of the economy becomes higher. It also justifies the need for taxing agricultural produce based on the fact that agriculture also benefits from government spending in the form of infrastructure investments, provision of low interest loans, fertilizer subsidies, and public research and development efforts (World Bank 2007a).

In the case of Punjab, land revenue was abolished and replaced by an Agriculture Income Tax (AIT) in 1996 (Gazdar 2011). These rates were amended during 1998, and the last amendment was made during 2003. However, even the AIT raises little revenue. Besides exempting up to 12.5 acres of irrigated land from tax, which does provide some relief to poor farmers, the AIT currently taxes all acreage of farmland on extremely low rates, which do not differentiate between the levels of productivity of individual plots of land. For example, for a parcel of 50 irrigated acres of land, the tax rate is Rs 250 for each acre, despite the actual productivity and income variance across this land. As a result, real per capita collections fell by two-thirds from 1999 to 2005 in Punjab (Shafique 2008). The lack of accurate, timely, and detailed data on ownership of land, crop yield, and input prices leads to a situation where there is little basis to establish AIT liabilities accurately.

While the need for a direct agricultural tax based on incomes was meant to be a progressive measure, aiming to prevent burdening poor subsistence farmers, the World Bank has begun pointing out that about 85 per cent of all rural landowners in the province are currently outside the existing tax base. It has identified the prospect of generating further revenue by reducing the exemption in Punjab from 12.5 acres to 5 acres, or else to have no exemption and instead charge a nominal (flat) amount for small farms. The World Bank has even highlighted the possibility of taxing the rental income of share-croppers, but recognizes that there is little evidence that information on rental values and tenants is readily available (Gazdar 2011; World Bank 2007a).

Provincial-level officials in Lahore, from across different departments, how-ever, took oppositional positions concerning World Bank assertions. Some of them felt that 12.5 acres of irrigated land is in fact well above subsistence, and there remains ample room for the taxation of farmers who have even less land. Others said this was not the case, and attempts at taxing farmers are not justifiable at all. The Executive District officer (EDO) for Agriculture in Umerkot, for instance, was not in favour of the agricultural tax, since he felt that the lack of water and the depleting soil quality already constrained the potential for maximizing yield, and any further pressure on farmers would cause agricultural production to decline.

One official in the Law and Parliamentary Affairs Department argued that the provincial government is not justified in placing an increased tax burden on farmers, since it does not give them the subsidy support provided in the developed world. He blamed the World Bank and the WTO for being unable to compel developed countries to open up their agricultural markets to countries like Pakistan, and for continuing to pressure one-sided removal of support to agriculture in the developing world, which was exacerbating problems like food insecurity. Despite the fact that agricultural trade liberalization is shown to be threatening the food and livelihood security of millions of poor farmers by lowering the prices of crops produced by them due to imported subsidized agricultural crops (MHHDC 2009), the official was arguing not specifically to bolster poor farmers, but rather to lessen the burden of taxation on the farming community in general. This rationale is justifiable form the viewpoint of an elite-led growth strategy, which has been adopted by many countries, including Pakistan, beginning at the time of the Green Revolution several decades ago. However, given the necessity to increase the overall revenue generation capacity of the cash-strapped state, and the need to supplement poorer farmers, a blanket tax amnesty for the farming community seems less tenable.

It is not surprising that large landowners also present similar arguments, citing for example, the high cost of inputs, fluctuating market prices, or the diminishing productivity of land, as reasons to not pay agricultural taxes. But all these problems are being confronted by poorer farmers as well, including small landowners and sharecroppers. The lacklustre implementation of the AIT in Pakistan again demonstrates how powerful landowners are able to block policies that go against their economic interest. They have been successful in their efforts with support from government officials who have also shown a lukewarm response to using AIT to tax actual farm incomes. Instead, government officials continue charging a meagre amount based on landholdings. Due to the reluctance of large landowners to pay a greater share of tax revenue to the government on the basis of their incomes derived from agriculture, the capacity of the state to invest in improved infrastructure (farm to market roads, irrigation, etc.) to benefit the farming community as a whole continues to diminish. Poorer farmers are the ones who face the brunt of this diminishing state capacity to facilitate agricultural growth. It is poor farmers who also lack the leverage with state institutions to access the increasingly scare state resources available, which continue to be monopolized by the landed rural elite.

c. Ineffective agricultural extension and veterinary services for poor farmers

State reluctance to generate needed resources by progressively charging those who can afford to pay is prevalent, and can readily be discerned in agricultural extension and veterinary services departments as well. There are distinct pro-vincial government departments in both Sindh and Punjab for agriculture and livestock and dairy development, which aim to perform a variety of functions. Agricultural departments have research units aiming to collect data, conduct research, and undertake agricultural marketing. Similarly, livestock departments aim to boost production through breed improvement programmes, focusing on animal health and other aspects pertaining to livestock development. In this sub-section, I have chosen to focus particularly on ongoing agricultural and veterinary extension services that directly aim to improve farming and livestock rearing techniques,[31] for not only increasing production efficiency, but bettering the lives of the general farming community. The findings emerging from the analysis below will indicate a dearth of targeted veterinary extension services aiming to directly benefit poor farmers, and conversely how a recent attempt to waive 'user-fees'[32] for veterinary services is instead undermining the interests of poorer farmers.

Agricultural extension workers are meant to provide services to farmers as per their actual needs, but they are criticized for passing on irrelevant, outdated and ill-timed prescriptions, formulated by bureaucrats, who are not in touch with ground realities (Shafique 2008). This critique has been accompanied by the increasing trend towards the privatization of agricultural support services, endorsed by donor prescriptions of curbing public spending expenditures. By dismantling the public monopoly on supplying inputs and services to farmers, it was assumed that a transition from subsistence to commercial agriculture in Pakistan could be facilitated, as private business interests and competition would help improve agricultural performance and increase the technical knowhow of the farming community, including poor farmers (Shafique 2008; World Bank 2007a).

Aiming to assess the veracity of such claims, I met a range of officials within the Agriculture Extension Wing of the Agriculture department heading the soil conservation, forestry, water management, and fertilizer sections, as well as field assistants responsible for the supervision of demonstration plots and undertaking the mobilization of farmers. Most of these officials admitted that they lacked any specific mandate to interact with agricultural labourers or sharecroppers. Instead, they focus primarily on people cultivating their own land or that on lease, since they are the ones with the power to make decisions concerning what to grow and how. The fact that there are no female officers for extension services in the district was also said to cause difficulties in reaching out to women involved in agriculture.

Some of the field assistants, who are tasked to work directly with farmers, described smaller farmers to be hard-working, but in need of more support and advice to improve their yields. One of the field assistants gave the example of poorer farmers often overusing area as it is cheaper, instead of using it in conjunction with potassium or other fertilizers, which are vital for soil nourishment. Another field assistant, however, dismissed the need for making an extra effort to impart relevant advice to poor farmers, based on the rationale that these farmers often lack the resources to pay for the required additional input costs.

A majority of field assistants admitted that government programmes often fail to provide resources relevant to the needs of poor farmers. For instance, they mentioned a provincial food security program, being implemented in Gujrat district during 2008 and 2009, which had provided subsidized implements to help boost food production. These implements included mechanized seed drills, seed graders to sift through broken seeds, and pesticide spray machines. However, only the spray machine could be provided to small farmers, since ownership of a 5-acre plot was a precondition to get the other implements. Farmers who do not own tractors could not even use seed drills.

A lengthy discussion with a group of agricultural department extension officials, including field assistants, led to an admission about the preconditions of the said food security scheme being irrelevant for poor farmers, who generally have fewer than 2 or 3 acres of land. These extension department officers however merely shifted the blame for the inappropriate scheme onto senior officials making ad hoc policies at the provincial level.

I also met livestock department officers in Kotput village in Gujrat, who were implementing a livestock vaccination campaign. It was interesting to note that the vaccination team of the livestock department had set up camp at the *dera*[33] of one of the most influential landowners in the village (Haji *Sahib*), and the entire vaccination campaign was being coordinated from there. I was informed that this particular spot was made available to them regularly, whenever they visited the village. It was an informal arrangement, which suited the needs of both the vaccination team and Haji *Sahib*. The fact that a government service was being coordinated from his *dera* furthered the prestige of this landowner within the community, while the vaccination team had a central place to coordinate their efforts, rest, and drink tea. Haji *Sahib* informed me that he had been in touch with the livestock department's field officers to coordinate the campaign beforehand. Moreover, Haji *Sahib* was also offering advice to the field officials concerning where to go next. Regular visitors to this *dera*, which included other prominent landowners, were also coordinating with the vaccination team, to ensure that their livestock would not be neglected during the visit. Except for landless servants working for larger landlords, I did not notice poorer farmers approach the *dera* to coordinate vaccination by livestock officials. This was despite the fact that the field officer in charge of the vaccination campaign had informed me that there was no longer a fee being charged for vaccinations.

Up until the recent past, the Punjab government used to charge a fee from livestock owners with more than five livestock but now the minimum amount of livestock to be vaccinated without a fee had been increased (to 20 livestock). This policy revision was due to pressure exerted by numerous farmers in the vicinity, who had good contacts with local politicians and within the bureaucracy itself. One field officer said that the policy reversal had made their work easier, since previously farmers with more than five livestock would object to paying a nominal fee ranging from Rs 1.50 to Rs 6.50 for different vaccinations. However, this fee waiver implies more cost for the government, even though it is not really a pro-poor subsidy given that it is being availed by farmers with between five and 20 livestock, who are not really very poor farmers.

The fact that such a policy reversal would decrease revenues available within the department, and have a corresponding adverse impact on the amount of veterinary services that can be provided, was not something that the field officer himself seemed too concerned about. A more senior official justified using flat, rather than differentiated user charges. He argued that providing livestock officers the discretion to determine whether or not, and at what rate, a farmer is to be charged for a particular service would merely provide more rent seeking opportunities. However, given the prevalent institutional bias towards larger landowners, as well as the observed pattern of more affluent farmers being given priority in terms of service delivery by the veterinary field officers, as described above, it seemed more likely that poorer farmers would be the ones excluded from availing livestock vaccination services, if the veterinary department has to curb the programme in the future due to resource constraints.

Rather than utilizing progressive policies to ensure revenue generation without placing an undue burden on the poor, the vaccination fee charging policy reversal indicated the opposite trend of widening the scope of subsidization to benefit middle income farmers, even if its ultimate outcome reduces the institutional capacity of the state provider to extend outreach or increase the scope of its services.

Instead of offering suggestions whereby poor farmers could have been subsidized more selectively, while generating sufficient revenues by charging those farmers who can afford to pay for the vaccination services, some officials even made the case for abolishing small fees altogether, based on the rationale that the cost of attempting recovery is larger than the amount recovered. Yet the abolition of such fees seems unlikely given the resource constraints faced by the public sector in Pakistan, as well as the pressure by the World Bank to charge user fees to help finance government spending.

Besides the problem of diminishing revenues available for essential departmental services from reaching poor farmers, the broader issue of agricultural subsidies has also become a source of contention, as the following sub-section indicates.

d. A closer look at agricultural subsidies

The issue of agricultural subsidies remains contested between donors and the Pakistani government. The World Bank argues that subsidies increase public expenditure, whereas the government sees it necessary for boosting agricultural development. Substantial liberalization took place from the mid 1980s to the early 1990s, however, greatly reducing explicit tariffs and taxes, as well as government direct interventions in markets for most agricultural products. Yet, the government continues to subsidize some agricultural inputs such as fertilizers, as well as controlling prices for major food crops such as wheat.

As the broader debate concerning the merits or demerits of subsidization on agricultural growth in general is beyond the scope of this book; the following section mainly attempts to draw attention to how corrupt practices and ill-devised government subsidy schemes continue to deprive poor farmers of much needed government support.

Many poor farmers across the districts I visited complained about shortages and the escalating prices of agricultural inputs, which were making it increasingly difficult for them to meet household expenditures. It was surprising to hear these complaints, given that the government subsidies state-managed fertilizer companies in order to provide urea to farmers across the country at well below global import rates. When I raised this issue with some analysts and civil society organizations in Lahore, they immediately drew my attention to a major recent scandal involving a powerful group of federal ministers, elected officials, and senior civil servants, who had allegedly embezzled Rs 300 billion through the illegal dumping, smuggling, and black-marketing of fertiliser imported by the government to subsidize farmers in the country.

In December 2011, a major scandal broke out involving two former Ministers, a Senator, as well as a large number of senior officials from the Industries Ministry. As a result, the state-owned Trading Corporation of Pakistan, and the National Fertilizer Marketing Ltd (NFML) stopped delivering subsidized urea to Pakistani farmers. Over 50 per cent of the imported urea, some 500,000 tons, did not reach its destination, since almost 600 fake dealers and agencies were created and given authorization to distribute subsidized fertiliser with the support of corrupt officials and public representatives. Once the fake dealers were able to procure the cheap fertiliser, they dumped it into storage sites across the country and waited for the resultant artificial shortage of urea to drive up local prices, forcing farmers to pay up to double of what should have been the subsidized price (Rana 2011).

Such instances of corruption are not isolated, but in fact recurrent, and they provide justification for entities like the World Bank to lessen government interference and increase liberalization of the agricultural sector. While corruption of this sort has an overarching detrimental impact on the farming community in general, poorer farmers, including sharecroppers who have to bear half the burden of input costs, feel the brunt of such corrupt practices. The increasing costs of cultivation push poor farming households into increased debt, hunger, and other forms of deprivation, which the resource constrained government is ill equipped to address given that it wastes its scarce resources on schemes whose benefits do not reach the intended beneficiaries.

The government also has a regulatory mechanism in place to provide an assured price for farmers who produce wheat, and for ensuring that wheat is supplied onwards to flour mills and consumers at reasonable prices. Wheat actually plays a central role in Pakistan's food economy, both in terms of production and consumption. Because of its importance, successive governments of Pakistan since independence in 1947 have intervened heavily in wheat markets, procuring wheat at administratively set prices (AHRC 2009). Some steps were taken towards liberalization of wheat markets from the late 1980s to 2000, alongside other crops such as rice. However, after consecutive relatively poor wheat harvests from 2002 to 2004, which led to high market prices for wheat, the federal government, as well as the government of Punjab, began supporting fixed procurement prices again (Gazdar 2011).

Conversely, entities like the World Bank argue that the government should allow the free market to operate for wheat. A manager of a project funded by the United Kingdom's Department for International Development

(DFID), working on a study concerning food security blamed government involvement in the process of wheat procurement for encouraging rent seeking. He also dismissed the need for keeping more than a three week food supply within the country, which was deemed sufficient time for importing grains in case of an emergency. 'Why have two spares in a car?' he queried. Yet, the provincial government's resolve to control the wheat procurement process has persisted.

However, the benefits of the government controlled support price for wheat are not reaching all farmers equally. Analysis of data taken from the Pakistan Integrated Household Survey (PIHS) 2001–2 confirms that wheat sales are highly concentrated, where the top 10 per cent of wheat farmers, in terms of sales (derived from their larger landholdings), account for 47 per cent of total wheat sales (cited in Dorosh and Salam 2007). Despite quantitative documentation of such procurement biases, there is less information available concerning implementation loopholes, which allow larger, instead of smaller and poorer, farmers to avail the government-controlled wheat prices.

I focused on exploring the wheat control pricing issue within the Gujrat district in Punjab to understand the phenomenon of biased wheat procurements. For this purpose, I repeatedly visited a government wheat collection depot in Mangowal,[34] while it was engaged in the process of wheat procurement. Food Department officials in Mangowal store wheat at an open-air wheat collection depot located close to a cluster of villages, besides more centralized warehouse buildings with greater storage capacity. Officials from the agricultural, food and land revenue department are tasked to monitor the wheat procurement process. These officials cross-check each other's work to ensure that landowners, rather than commission agents, are selling wheat to the government, and to ensure that the wheat being bought is of good quality and farmers are being remunerated on time. Once the wheat procurement has taken place, flour mills can buy wheat from the government depot, and from its warehouses, as per the specific quotas allocated to them.

The Food Department procurement in Gujrat for 2011 began at the end of April and went on until June. The Mangowal wheat depot was meant to achieve the government set target of procuring 80,000 bags weighing 40 kg each. During 2011, the government had set the official price of Rs 950 per 40 kg bag, with an additional Rs 7.50 to compensate for delivery charges per bag. The government procurement process does not take the responsibility of picking up produce from the farm-gate. The additional Rs 7.50 per bag given to farmers for delivery charges, however, does not cover the actual cost of labour and transportation, which can range from between Rs 20 per bag to Rs 30 per bag during peak procurement season. An inadequate procurement process was identified as one of the reasons why poor farmers need to involve intermediaries such as commission agents in the sale of their crops. These commission agents pick up the wheat from the farm-gate. Even if they pay farmers Rs 30 to Rs 50 less than the government set control price, they make payments in cash, and then these agents take the responsibility of transporting the wheat away to sell onwards to either the government wheat depots, or else to flour mills, which often need more flour than the government allocated quotas.

Commission agents further pointed out how the government procurement process is very cumbersome, which serves as a disincentive for poorer farmers who do not have time to leave their crops around harvest time to go fulfil the required government procedures. In order to sell their wheat to the government, farmers need to deposit money to procure official sacks required to sell grain to the government procurement centres, wait in lines to secure approval from land revenue officials, open a bank account, and then to wait for money to be deposited days after they have sold their crop. Poor farmers do not have storage facilities and they do not want to leave their wheat out in the fields for days without insurance, while in the process of dealing with official procurement procedures. All these factors combine to compel poor farmers who lack the holding power and resources from availing themselves of the government-offered price for wheat procurement.

Discussions with government officials around the exact procedures concerning wheat purchasing further revealed why the existing procurement process tended to favour larger rather than smaller farmers. Food department officials informed me that they did not purchase fewer than 10 bags of wheat from an individual farmer. However, no one had actually bought 10 bags during the 2011 procurement process in Mangowal. The least amount of wheat sold to the Mangowal wheat depot was around 25 bags, and on average farmers had brought in 100 bags each, which implies that they were growing wheat on over 12 acres of land as per government estimates of wheat production per acre. Moreover, the *patwari* for Kotpur village (also in Mangowal) informed me that fewer than a dozen villagers had been able to sell their wheat to the Food Department during 2011, which obviously implied that numerous farmers were unable to benefit from the government controlled price for wheat. Moreover, Haji *Sahib*, the prominent landowner in the Kotpur village, further admitted that government officials expedite the process of procurement, and perhaps are less particular in terms of rejecting the grain due to quality issues, when they are dealing with farmers of influence, in comparison to poorer farmers.

While most food department officials in Gujrat claimed that their procurement process had been fairly transparent, some of them acknowledged that corruption and biases in the wheat procurement process did take place, but they claimed that such problems were more rampant in southern Punjab where much larger landowners reside.

The micro-level analysis above also illustrates why large landowners find it easier to benefit from wheat control prices than poorer farmers. The fact that many big landowners are also politicians, or else enjoy close ties to politicians and government officials, explains resistance within the Pakistani state itself to heed donor advice to abandon controlling wheat pricing, since the benefits of this fixed pricing are largely availed by larger landowners. Yet, the rationale for preserving this wheat price regulation employs pro-poor rhetoric, such as claiming to help 'poor farmers against market volatility' (Government of Pakistan 2003: 48). How the scheme is actually designed, however, does not achieve such aspirations.

e. Government officials' attitudes and practices concerning poor farmers

Besides the evident lack of specific mechanisms to ensure the inclusion of poor farmers within major government schemes to bolster agricultural development, the evident biased attitudes and behaviour of key government officials (including a senior bureaucrat/decision maker, *patwaris*, and police officials) concerning issues of direct relevance to poor farmers are also instructive.

A retired bureaucrat having served in the highest levels of key government departments, including finance and agriculture, and who had been closely involved in loan negotiations with donors like the World Bank, was dismissive of developmental economists' concerns regarding asymmetrical markets, which tend to charge poor farmers more for inputs and lessen the price of their outputs in comparison to larger landowners. Despite being familiar with evidence provided by the National Human Development Report for Pakistan pointing to existence market distortions in rural Pakistan[35] (Hussain *et al.* 2003), this former bureaucrat maintained that in large parts of Punjab where agriculture was well entrenched, such distortions do not occur. He did admit that such distortions may exist in remote rural areas, but he claimed that the contribution of these remoter agricultural areas to the agricultural production process was not significant. He did not pay much heed to the impact of such distortions on the lives of the poor farmers, nor the fact that many of the 'remote' rural areas in the country have sizeable populations given the overall population of the country. While conceding to the need for crop insurance, he pointed out that this was a difficult proposition since a majority of farmers were still poor, so they wouldn't be willing to pay for another expense. He was not in favour of the government incurring this cost, which would be another subsidy, which the donor agencies would disapprove of as well. He was also dismissive of the need for setting decent minimum wages, saying that such notions inhibit Pakistan from using cheap labour to boost productivity. This myopic mindset of a high-level bureaucrat was apathetic towards the human costs associated with exploitative labour, borne by poor rural men and women struggling to ensure household survival.

I spent a lot of time with several junior land revenue officials, called *patwaris*. The *patwari* is the most junior official in the Board of Revenue's administration. Despite little formal support (such as provision of a vehicle or even an office), the *patwari* has numerous important duties, including maintenance of land records and determining land revenue liabilities. A *patwari* also provides certification of landownership needed to secure agricultural loans. Often a *patwari* is also called to court to attend hearings and provide evidence in cases of land disputes (Gazdar 2011).

All the *patwaris* I met, complained that they are underpaid, and do not even get any transport facilities, despite their need to travel for official purposes. The fact that the duties assigned to *patwaris* are much more significant than their official rank and remuneration does make them prone to corruption (Gupta 1995). Due to their important role in land revenue management, *patwaris* are very sought after by the local elite, who receive favours in return for patronage, often to the disadvantage of poorer farmers (Malik 2009). One apparent means of favouring landowners is for *patwaris* to avoid mentioning the name of sharecroppers on land cultivation records, which enables landowners to have a much stronger bargaining position and legal status in comparison to their tenants. While the *patwaris* I interviewed did not admit having any special ties with larger landowners, the evident patronage of *patwaris* by landowners was not hard to establish. Most of the *patwaris* I met were working from workspaces belonging to local landowners. Once I became friendly with some of these *patwaris*, they admitted that revenue accounts can also be easily altered to undervalue the productivity of land, so landowners do not have to pay high taxes. By helping larger landowners evade taxation, these government officials, in turn, undermine the capacity of their own public institutions to generate sufficient funds to function properly, and provide services to a larger number of farmers, including those who are poor.

The problem of officials like *patwaris* being prone to corruption was acknowledged by some senior officials as well. A senior provincial official in the Law and Parliamentary Affairs Department (based in Lahore) pointed out that it is not uncommon for aspiring *patwaris* to pay up to Rs 4,000,000 to be successfully recruited, a payment which is then extorted, especially from poorer farmers who do not have the power to resist their monetary demands in order to get their necessary work done. Gupta (1995) has undertaken an intriguing Foucauldian analysis of this phenomenon, whereby he views the corruption of local officials such as *patwaris*, as the means by which the otherwise under-resourced state manifests itself in far-flung areas of developing countries such as India. The corruption of government officials like *patwaris* can also take on other forms than bribe taking. A *patwari* who was one my key informants in Gujrat, for example, had set up a side-business while working full-time in his official capacity. While not from a landed family himself, the *patwari's* father had worked as a labourer overseas, and the money the father brought back was used by the *patwari* and his younger brother to set up an electronic goods shop. Initially the *patwari* mentioned that it was primarily his younger brother running their shop, but during multiple visits, I would find him sitting at the shop instead, and some of the villagers even came to see him there concerning land related issues. The *patwari* justified his involvement in a side-business, as a necessity for him to meet his personal household expenditures, given the inadequate remuneration offered for his full-time job as a *patwari*.

While most of the villagers I spoke with did not accuse this particular *patwari* of corruption, many of them did complain that their *patwari* was hardly present at the village. Some of the villagers even accused the *patwari* of having sub-let his responsibilities to a *patwari* of an adjoining village. While I could not validate this accusation with the *patwari*, without endangering the villagers, I did on several occasions ask him about particular instances in the village, such as a recent fight over land, which he did not seem to have any information about despite being assigned to this particular revenue circle for four years. Given the fact that *patwaris* usually have extensive knowledge about villages in their designated jurisdiction, it was clear that this particular *patwari* was not paying enough attention to his area of administration.

Due to different reasons including patronage based biases towards land-owners, neglect of their official duties, or just the difficulty of managing a vast amount of land records manually, numerous inaccuracies exist in agricultural land records. These inaccuracies often spark land-related disputes, and many of them can take a violent turn, leading to police involvement and lengthy court cases, which poorer farmers especially cannot afford to commit the time or resources to. The role of police stations in impoverished rural communities is vital, since they are often the only tangible manifestation of state authority in such remote areas. However, the influence of larger landowners over police stations is also well recognized, whereby these landowners can often exert control over poor farmers by having them implicated in false cases, or preventing the police from officially registering their complaints (Malik 2009).

Several government officials verbally reiterated the need for government officials to be responsive to the challenges confronting poor farmers, despites the evident dearth of institutional arrangements and policy measures to ensure that the interests of poor farmers are taken into account. I also found evidence of biased attitudes and practices prevailing among government officials, including those in the police.

During an interview of the most senior police official (a deputy super-intendent) in the Gujrat district, for instance, I was informed of the presence of several land grabbing groups within the district who take possession of disputed land and try to forcibly evict people inhabiting this land through the use of force. He admitted that politicians of the area often provide protection to these land grabbing mafias, and that poorer farmers fall prey to such ten-dencies more easily. Yet while acknowledging the problem of land grabbing, and even implicating politicians in the process, the police official was careful to distance his own department from the problem. The same police official, while discussing the issue of land disputes between tenants and landowners, stressed that it was actually landowners who would often be exploited by tenants who did not pay their rents on time, instead of tenants being victimized by landowners. In the case of Gujrat, many tenants are not disempowered sharecroppers but instead fixed-rent tenants, often farming land obtained from several smaller landowners. It was thus not inconceivable for them to try and exert themselves unfairly against small landowners. It was surprising, how-ever, to hear other government officials in Sindh saying similar things, despite the fact that landholdings in districts like Badin and Umerkot are much more inequitable. The attitudes of such officials are indicative of prevailing elite biases towards the poor, which often stereotype the poor as being lazy or prone to crime, which in turn is cited as a justification of their underprivileged position in society (Gilens 1999; Reis and Moore 2005).

The illustrative examples above provide some indication of the pervasive negative attitudes and problematic practices concerning poorer farmers, which continue to exist within prominent government institutions and among its key officials.

Conclusions

Despite constitutional guarantees to protect the rights of all citizens, a broad range of state institutions tend to preserve and protect the interests of large landowners rather than addressing the underlying causes of widespread deprivation of the rural populace, which comprise poor farmers.

The PPP government, during its last tenure (2008–13), did make an attempt with the passage of the Women Protection Bill 2011 to facilitate women ownership of land. However, this legislative attempt was problematic due to its myopic and primarily punitive approach towards women's lack of landownership, which ignores both the sociocultural compulsions that prevent women from claiming their land rights, and therefore did not include supplemental measures required for encouraging them to register complaints in case of their inheritance being appropriated by male family members, or to support them in taking such grievances to court.

Many of the other recent legislative attempts to instigate agricultural development are more in line with market-led donor development strategies, such as the promotion of plant breeder rights, or corporatization of agriculture, which are consistent with the broader donor-led agenda to increase the liberalization of the agricultural sector. Such policies provide opportunities for larger landowners to increase yields by using more expensive inputs, or by leasing out their lands to corporate farms, but they have adverse impacts on the lives of poorer farmers by making inputs more expensive and compounding land scarcity (see Chapter 5 for more details).

After decades of inaction, a proposal for land reforms was recently formulated by the urban-based MQM, which presented a draft bill to this effect in parliament. However, this attempt was viewed as an opportunistic political move, and was promptly dismissed by other legislators.

The judiciary has not played a proactive role in altering the status quo of landownership patterns either. In fact, the Supreme Court's decision over two decades ago, which declared land reforms 'Un-Islamic', has not been revoked despite recent public interest litigation calling for a review of this judgment.

Adopting a less radical approach than redistribution, the Benazir Landless *Hari* Scheme has aimed to provide state land to landless farmers, but this scheme also faced difficulties in empowering deserving candidates, especially women who were intended to be the primary beneficiaries. Moreover, the

NGOs working to implement this scheme have remained unable to help the poor beneficiaries of the scheme if they have not been given a fertile enough plot of land, or if any of their legal documents were found to be incomplete. The impact of this particular scheme, which offered a limited amount of state land for redistribution, to change the underlying structures of patronage, or to even exclusively benefit genuinely landless households, has remained questionable.

Under donor pressure to achieve better revenue collection, the government began implementing flat rate charges in both the agriculture and irrigation sectors. However, these revenue generation reforms have shifted the burden uniformly to all farmers, rather than trying to make the state taxation mechanism more progressive by charging differentiated rates for poor farmers and larger landowners. The case of the livestock department revising its fee charging policy upwards under pressure from medium-sized farmers, unwilling to pay for vaccination services, also reveals the continued resistance to instituting progressive revenue generation attempts. Furthermore, such measures hurt poorer farmers indirectly as well, since they in turn experience an ever diminishing capacity of public sector institutions to improve the efficiency and capacity of vital infrastructure and provision of essential services. Given the prevailing circumstances, it is in fact larger landowners who continue to utilize the scare state resources available before they can reach poorer farmers.

While the state continues to resist donor pressure to stop subsidizing agriculture, existing agricultural subsidies are subjected to both corruption as well as elite-capture, by larger landowners. While the stated objectives of schemes such as the wheat procurement process are predicated on altruistic goals (of ensuring national food security or supporting poor farmers), corruption and design flaws within the schemes result in their benefits being availed by corrupt officials, larger landowners, or middlemen such as commission agents instead of poorer farmers.

Analysis of several relevant state institutions also indicates that there are no specific incentives for poor farmers, especially sharecroppers or agricultural labourers (both men and women). Instead, the very officials who are in a position to help them often view poor farmers with bias and suspicion. In short, tokenistic measures launched by successive governments have failed to adequately protect the rights of poor farmers, including landless tenants and agricultural workers, from varying levels of exploitation and depravation.

While this chapter has focused on ascertaining impacts of various state-led institutions and policies on poor farmers, the next chapter focuses on policies and programmes being directly supported by donors to assess whether they offer better options to address the needs of poor farmers.

Notes

1 *Dera* in the Punjabi and Seraiki language literally means a camp or settlement, but in an agrarian context it refers to a common area, designated on the property of landowners, where they can socialize or rest.

2 Since my car and the driver were near enough to have prevented any confusion when Arif made the comment, his comment seemed more for the sake of effect, perhaps to indicate to me that he was also wealthy enough to own a Rs 1.5 million car and hire a driver, even though he was cultivating someone else's land.

3 The term *haji* sahib is often used to refer to people who have performed the Muslim ritual of *haj*.

4 It is not uncommon for women and children of indebted sharecroppers to also be working to help pay off their family loans. Unpaid loans can also be transferred from one generation to the next, which results in intergenerational debt bondage.

5 The PML-N, with Nawaz Sharif as the Prime Minister, was elected to power first in 1990, but the president dismissed this government prematurely in 1993. Thereafter, PML-N got re-elected to power in 1997 but General Musharraf again overthrew it, via a military coup, in 1999.

6 This term commonly refers to Arab countries bordering the Persian Gulf, including the UAE.

7 Pakistan became a signatory to TRIPs in 1995.

8 This is a Latin term, literally meaning something which is of its own kind, or has its own unique characteristics. In this case, the term is used to imply that governments may develop their own particular means/legislation to ensure patent rights.

9 Programme managers form a leading national NGO, a regional NGO working across South Asia, and an international NGO were asked specific questions on this topic during their semi-structured interviews.

10 The MQM was formed in the mid 1980s to represent the interests of Urdu-speaking *mohajirs* or migrants who came and settled in Pakistan after the partition of the Indian subcontinent in 1947. The MQM has become a prominent political party in the country, due to its entrenchment in Karachi, and also in Hyderabad, in Sindh. The MQM leader is, however, in self-exile in London, and the party is alleged to control land mafias in the largest city of Pakistan, Karachi, which has an estimated population of about 18 million.

11 The Supreme Court in the 1990 Qazalbash *waqf* (religious endowment) case declared that resumption of land by the state and imposing land ceilings were 'unIslamic'. The argument made was that the religion has proscribed obligatory charity, but it does not place limits on the wealth of individuals, including on landownership.

12 Absentee landlordism implies an economic arrangement still prevalent in rural areas of Pakistan whereby owners of large parcels of land do not live on their land but instead rent it out or enter into sharecropping arrangements using intermediaries.

13 The said decision pertained to the landownership rights of a poor farmer in the Bahawalpur district of southern Punjab, who had three *kanals* of land transferred to a retired army Brigadier under a larger scheme which had allocated over 33,000 acres of provincial government land to the Army GHQ in 1993. The Punjab government seemed to have transferred this land without checking its ownership titles accurately. Hence, the brigadier got around 46 acres of land allocated to him, which included about three *kanals* belonging to the said poor farmer. The retired Brigadier contested the poor farmer's ownership in the High Court and the Supreme Court, both of which upheld the farmer's ownership rights.

14 While registered political parties, neither WPP nor the NP have any have managed to achieve electoral success to achieve any significant representation in any of the provincial assemblies or at the national level.

15 Customary practices such as *vani* or *swara*, which force girls to marry members of another kinship group to help settle feuds, were criminalized under the Anti-Women Practices Bill 2011.

16 As mentioned before, the land reforms in the late 1950s and the 1970s placed ownership ceilings on individual landholding, rather than that of families, and

many large landholders thus evaded resumption of their land by transferring their land into the names of their individual family members.

17 Being a distinct Muslim sect, *Shias* follow the *Jafari* school of religious law, offering slightly different rules for women's inheritance.

18 Due to being a British colony, the notion of 'common law' was introduced to the Indian subcontinent. Common law is essentially developed by judges through court decisions, as opposed to statutes adopted through the legislative process issued by the executive branch of government. Since the statute laws were not written afresh at the time of independence, common laws remain influential in Pakistan, but the country's status of being an Islamic republic has also enabled Islamic laws to be incorporated into the legal system.

19 While English common law is applicable to most commercial relations, Muslim personal laws are incorporated into the legal system in Pakistan to deal with personal matters such as marriage, divorce, inheritance, guardianship, etc.

20 A dowry is the cash or in-kind resources that a woman brings to a marriage. While there is no concept of dowry in Islam, the custom is widespread in Pakistan in practice. The Islamic equivalent is the dower which is to be provided by the husband to the wife when the marriage is contracted.

21 This instance provides an illustration of the fact that women themselves often reinforce patriarchal norms.

22 An alternative conflict resolution mechanism body set up at the local government level to help resolve local disputes, where no serious crime has yet been committed, through compromise instead of resorting to lengthy legal proceedings.

23 Somewhere between 60 and 75 per cent of the entire rural populace of Pakistan is estimated to be landless (Anwar *et al.* 2004; Malik 2010).

24 About 85% households own no land in Sindh (out of a population of approximately 40 million), which is the highest level of landlessness in comparison with all the provinces in the country (Anwar *et al.* 2004).

25 The government was meant to use a database developed by RSPs using poverty score cards, which rank people on the basis of asset assessment of individual households using a structured questionnaire. Based on verifiable poverty scores, households are then divided into different categories such as destitute; chronically poor; vulnerably poor, etc.

26 Cases of corruption have also been noticed in other government schemes for poor farmers. In September 2009, for instance, the government announced the Benazir Tractor Scheme which was supposed to award thousands of tractors to randomly selected small farmers across Pakistan using a computerized lottery programme. While the scheme was meant to only enrol farmers will fewer than 25 acres of land, many of the emerging 'winners' of this scheme owned thousands of acres of land, and included 48 relatives of a serving parliamentarian (Kugelman and Hathaway 2010).

27 Cropsharing arrangements between landlords and those tilling the land vary across and between different regions of the country. Sharecroppers can get between one-half and one-quarter of the crop depending on whether they pay a share of inputs or not. Often sharecroppers take advances from landowners to pay for these inputs, or to meet other costs, which are then deducted at the time of harvest. However due to lack of adequate documentation, the share of crops withheld from sharecroppers in lieu of loan repayments can be excessive and result in disputes.

28 According to recent estimates, 38 per cent of Pakistan's irrigated lands are waterlogged (Kamal 2009).

29 Service charge applied by the irrigation department to supply water to farmers.

30 There are two main harvest periods for growing different crops in Pakistan, known as the Kharif or autumn, and Rabi or spring.

31 Veterinary services are also being included in this analysis since livestock is often the most relevant asset for poor households, access to which is as, if not more, important to them as is access to land.

32 The term 'user fee' implies charging citizens, including the poor, a small fee to avail basic services like education for their children in government schools or to access public hospital facilities.

33 As explained earlier, a *dera* is a common area designated on the property of larger landowners, mostly for men to gather and socialize. In the context that this word is being used here, however, it is also important to note that larger landowners have more prominent *deras*, and sometimes play host to more significant events such as resolving community disputes or entertaining officials or other visitors to the village.

34 Mangowal is a union council in the Gujrat district, where one of the villages selected for this research was also located.

35 The mentioned report found that poor farmers have to pay more for inputs and get less for their outputs in comparison to larger farmers. This issue of 'asymmetrical markets', and the reason why it occurs, is discussed in more detail in Chapter 5.

5 Donor influence on agricultural development and implications for poor farmers

In addition to the influence of colonization on the formation of the post-colonial state in Pakistan, donor agencies continue to exert a major influence in shaping development paradigms adopted by the state, including those that have a direct impact on the lives of poor farmers. This chapter begins by identifying the broad range of donor policies and programmes that have relevance for poor farmers. It highlights policies and programmatic interventions endorsed by Poverty Reduction Strategy Paper of Pakistan to address the challenges of agricultural development and rural poverty. It also examines implications of the increasing reliance on market mechanisms and on encouraging multinational corporations to participate in agriculture. Specific donor advice, such as attempts to support the computerization of land records, using NGOs as intermediaries in capital and land markets, and efforts to promote corporate farming, are then examined in some detail. This chapter also describes operations at a leading corporate farm in the country, with regards to its impact on land access in surrounding rural areas, and the opportunities such a venture offers for poor farmers, including women involved in agriculture. In conclusion, an overall assessment of increasing reliance on market-based interventions endorsed by donor agencies like the World Bank to address the marginalizing of poor farmers is provided.

Donor policies and programmes for poor farmers

The influence of multilateral and bilateral donors like the World Bank or the United States Agency for International Development (USAID) in Pakistan's agricultural sector can be traced back to soon after independence in 1947. The World Bank played a vital role in the formulation of the Indus Basin Treaty between India and Pakistan in 1960.[1] Subsequently, it supported a range of irrigational dam projects, as well as being a proponent and supporter of the 'Green Revolution' in Pakistan in the late 1960s and 1970s. The so-called 'Green Revolution' aimed to support state-led attempts to boost agricultural growth through the use of mechanization, and the increased use of subsidized agricultural inputs. However, it was primarily large landowners who received government support, and subsequently enjoyed the benefits of the resulting growth (Niazi 2004). Conversely, the 'Green Revolution' served to increase costs of production for smaller farmers due to the increased reliance on expensive inputs and capital, rather than labour intensive agricultural production (Alavi 1973; Gazdar 2005).

It was not until the late 1980s that more intrusive donor interventions were launched under the rubric of structural adjustment, compelling the Pakistani government to curb the elaborate system of subsidies and funding of public agencies meant to support agricultural production (Schuler 2004). One of the rationales employed by the World Bank for lessening public expenditures in the agricultural sector was that their benefits do not reach poor farmers (World Bank 2007b). Instead, the World Bank argued that liberalized agricultural markets, and private sector efficiencies, would provide adequate opportunities for inclusive agricultural growth under the WTO regime.

Research indicates that poor farmers often lack access to the required resources, knowledge, and technology needed to avail themselves of market opportunities or to meet export standards (Khan and Yusuf 2004). Prominent bilateral donors like the USAID have thus introduced certification programmes to help Pakistani agricultural products meet international standards (including environmental impact, use of chemicals, and worker safety), and enable growers to increase exports in new markets. Alongside a dearth of information concerning the overall success of such supplemental measures to help overcome impediments facing poor farmers, I instead found the above-mentioned programme being implemented at a large corporate farm in southern Punjab, the implications of which are discussed in detail below.

Bilateral donors like USAID closely align their agricultural development programmes with World Bank strategies, instead of paying attention to underlying causes of rural inequality such as inequitable landownership patters. This is despite advice offered by a USAID policy brief itself, which cautioned that religious extremists, including the Taliban, were aiming to garner support across rural Pakistan, by capitalizing on the prevalent disgruntlement over inequitable landownership (USAID 2010). This particular document suggested undermining extremist militants by promoting inclusive economic growth in rural areas, as well as providing opportunities for land access to the landless poor, including women headed households. Yet, until the time that this research was being conducted, no programmatic interventions had been initiated by USAID to help implement the above policy recommendations. Instead, millions of dollars had been committed by USAID for agricultural support programmes focused on increasing the profitability and incomes of medium-sized to larger-sized agricultural enterprises, and encouraging them to boost agricultural exports.[2] When I contacted the author of the above-mentioned USAID policy brief in 2012, he confirmed that the policy advice concerning USAID-supported initiatives to alter the existing landholding patterns in Pakistan had not been implemented through any concrete programmatic interventions.

The persistent lack of effective measures to help poor farmers seems surprising given the rhetorical emphasis of agencies like the World Bank to not only induce growth but also fight poverty. Particularly with the advent of the 'post-Washington Consensus'[3] within the World Bank, countries like Pakistan have been encouraged to formulate country-specific Poverty Reduction Strategy Papers (PRSP), which specifically aim to reflect the aspirations of the poor, and to increase the sense of country ownership of international development policies, by adopting a participatory approach to poverty reduction (Ali 2014; Oxfam International 2004). The PRSP process is meant to help governments of developing countries articulate a comprehensive strategy aiming to create a vital link between national public actions, donor support, and development outcomes, and to serve as the basis for channelling all foreign aid to poor countries (World Bank 2000).

However, the extent to which PRSPs are able to adopt participatory principles and represent the aspirations of the poor remains a controversial issue in many developing countries, including Pakistan. Although governments of developing countries were tasked to prepare the PRSP documents themselves, they had to get approval from the IMF and World Bank before concessional development funding would be approved. This led most developing countries to formulate PRSP documents proposing very similar interventions to those endorsed by the IMF and World Bank, such as facilitating market mechanisms in order to boost growth and alleviate poverty (Forster and Schnell 2003; Oxfam International 2004).

Although PRSP documents are meant to provide a poverty reduction strategy for a three-year period, and need to be revised subsequently, only two PRSP documents had been prepared by Pakistan at the time of writing this book; PRSP 1 in 2003 and PRSP 2 in 2008 (Government of Pakistan 2003, 2008).

Pakistan's PRSP 1 highlights the importance of the rural economy for poverty reduction and sustained economic growth. It recognizes that land is at the heart of agriculture and the rural economy in Pakistan, and that landownership and administration issues are of key importance. Despite acknowledging how the highly skewed distribution of landownership in rural areas has a substantive negative impact on agriculture productivity, the PRSP 1 for Pakistan simultaneously cites the counter argument that major land redistribution reforms merely lead to further fragmentation of landholdings and thus adversely impact productivity (Government of Pakistan 2003).

The World Bank also commissioned a research study to compare farming practices on tenanted and owned land in rural Pakistan, which concluded that greater tenure security instead of redistribution would be enough to increase land-specific investment on leased plots (Mansuri and Jacoby 2006). In turn, this assertion helped further deflect attention from the need for land redistribution, and instead provided impetus for the PRSP process to focus on creating a more efficient land record management system, hence encouraging the computerization of land records.[4]

The PRSP 1 of Pakistan also acknowledges the need for 'targeting a vast majority of small landholders to help improve productivity'. It further concedes to 'the major challenge to shift the focus of institutional credit support from landowners alone to the rural poor that includes landless tenants, agricultural artisans, women and other disadvantaged groups' (Government of Pakistan 2003: 47). But instead of focusing on underlying structural deficiencies, such as inequitable landownership patterns, which serve to marginalize these mentioned categories of the rural poor, the PRSP 1 places emphasis on the need for the provision of market-based strategies like micro-finance, which are not proven to be very effective in addressing the problems confronting poor farmers.

The PRSP 2 for Pakistan also acknowledges the inequitable distribution of land in rural areas being a major reason for rural poverty. But again, instead of aiming to provide land to smaller farmers, Pillar III of the PRSP 2 emphasizes the need for growth-led strategies such as increasing productivity and value addition in agriculture, development of new technologies, promoting the production and export of high value crops, and ensuring the availability of agricultural credit to farmers (Government of Pakistan 2008). No mention of the term 'land reforms' is evident in the entire PRSP 2 document. There is no acknowledgement of how poorer farmers tend to be excluded from the benefits of growth-led agricultural development strategies, despite ample evidence to this effect in the post-'Green Revolution' era of the prior decades (Alavi 1973). The PRSP 2 does not mention any specific steps to address the barriers faced by poor farmers in particular, such as their lack of access to export markets and their inability to avail government subsidies. Nor does the PRSP 2 propose any interventions to remove distortions in agricultural labour and capital markets, which work to the disadvantage of the poor (Ali 2014).

The PRSP 2 document only identifies the potential benefit of involving the landless poor in livestock or poultry schemes, or providing them micro-finance for other micro-entrepreneurship opportunities. It presents no concrete means by which to address the lingering problem of the inequality of land access. Instead, it endorses the need for encouraging trade and investment as the means to address the problem of rural poverty. In particular, it emphasizes the need for 'promotion of contract farming, [and] large-scale, vertically integrated agribusiness' (Government of Pakistan 2008: 86).

I asked a retired senior World Bank official, who had served in several important positions (including at the Country Director level in a Middle Eastern country), and had also worked as a consultant on the PRSP 2 document, why there was no mention of land reforms in the PRSP 2 for Pakistan. He explained how he had in fact been asked to help revise the PRSP 2 document, subsequent to unsatisfactory work by an earlier consultant. Given his more limited terms of reference, he worked with the existing draft document, which had also not taken up this issue. Besides avoiding the need for addressing the prevalent problem of landlessness among the rural poor, the manner in which this former World Bank employee had been hired to work on the PRSP document undermines claims of such policy instruments being vehicles for country-owned and participatory development strategies. Moreover, this former World Bank official maintained that he was personally against redistributive land reforms as well, since he felt that land fragmentation in Pakistan had led farm sizes to have already become very small, the management of which was not economically feasible. Instead, he reiterated his faith in the above-mentioned market-based policy prescriptions such as corporate farming and the need for the adoption of improved agricultural practices to help boost productivity.

While the World Bank places emphasis on the need for increasing outputs and enhancing incomes through adoption of agricultural technology,[5] actual adoption of such technologies is associated with capital and transaction costs that poor people are not able to afford. Donor agencies, such as the United Kingdom's Department for International Development (DFID), have recognized that poor farmers struggle to control production uncertainties, remain risk averse, and thus tend to benefit less than others from agricultural technologies as they stick with low risk, low return activities (DFIDb 2004). However, this same bilateral development agency continues to endorse the World Bank prescriptions of allowing the private sector to play an increasing role in the agricultural sector, as will be discussed in more detail later in this chapter.

Besides the PRSP 1 and 2, other World Bank documents, such as Pakistan's Rural Growth and Poverty Reduction Strategy, have called for further consolidation of PRSP 1 and 2 prescriptions for stimulating rural growth (World Bank 2007a). This has been done by the articulation of an integrated agricultural development strategy, which essentially seeks to reform input markets (seeds, fertilizer, extension), and factor markets (land, water, labour, and credit). A World Bank policy note on increasing agricultural productivity in the country has adopted a similar stance as well, by noting for example that 'sluggish growth in productivity has constrained farm income growth, limiting its potential for reducing poverty'. To address this problem, and to 'for accelerating broad-based agricultural growth', again the need for 'stimulating productivity growth through technology and innovation' are recommended (Ahmed and Gautam 2013: 1). The recommendations in this policy note also justify the need to address bias against agricultural exports to increase incomes and addressing land market rigidities to make land more accessible to smaller farmers.

Given the broader political economy of landownership, however, the pro-poor impact of such policies is by no means a foregone conclusion. The following subsections of this chapter draw attention to the implications of prominent agricultural development policies and interventions endorsed by donor agencies.

Markets, MNCs, and poor farmers

The World Bank emphasizes the need for liberalization and the market mechanism to secure agriculture's growth potential, and to generate the rural employment needed to alleviate rural poverty (World Bank 2007). Such assertions are also reiterated in the case of Pakistan, where, for instance, recent employment trends have been highlighted to prove how agriculture now provides the major source of employment for the female population, which, in fact, is increasing in comparison to the percentage of men working in agriculture declines. While men's employment in agriculture decreased from 43.4 per cent in 2000 to 35.2 per cent in 2008, the percentage of women involved in agriculture increased from 73.7 per cent to 73.8 per cent during the same time period (Government of Pakistan 2009b).

However, the data on increased female employment do not reveal the conditions under which women are working in agriculture. Critics, for instance, have pointed out how the liberalization of agriculture is forcing women to work longer hours, without adequate remuneration. There are also fears of women's health being increasingly compromised from exposure to hazardous pesticides (MHHDC 2003).

There is a dearth of donor-supported programmes addressing the underlying causes of women's exploitation in the agricultural sector. Donor approaches to agricultural development pay little attention to the socio-economic and political implications of inequitable landholding patterns, which create evident distortions in the agricultural labour and capital markets, particularly detrimental for poor farmers. A political economy approach to the issue of landownership is essential for understanding these agricultural market distortions, whereby an economic asset (land) reinforces other socio-economic and political asymmetries of power between poor farmers and large landholders.

Survey work done for the National Human Development Report (NHDR) for Pakistan demonstrates how poor farmers in Pakistan often end up paying a higher price for inputs than the landowners, and they also get lower prices when selling their produce (Hussain *et al.* 2003). These survey findings are based on a Pakistan Institute of Development Economics (PIDE) survey conducted for the NHDR in 2001, which estimated that the poor lose close to 37 per cent of their income due to severe market distortions. The NHDR also points out how asymmetrical tenure arrangements compel landless tenants/sharecroppers to pay a larger proportion of their farm produce to the large landowner as rent and payment for input costs, compared to non-poor landowners.[6] The PIDE survey for NHDR found that extremely poor (sharecropping) households are forced to keep only 39.59 per cent of their crop output for household consumption, compared to 48 per cent by non-cropsharing poor households and 54 per cent by non-poor households (Hussain *et al.* 2003: 62).

The NHDR itself does not provide details concerning the dynamics that result in such phenomenon. While the reasons for market distortions experienced by poor farmers seem varied, they cannot be understood without reference to a broader context of power relations between poor farmers, large landowners, and other intermediaries. My research tried to uncover some of the underlying compulsions which cause the above-cited market distortions. Discussions with landless sharecroppers in the Umerkot district of the province of Sindh provided some insights into how and why poorer sharecroppers are compelled to keep a smaller share of the crop. I also spoke with commission agents in different districts to explore why they give poor farmers lesser than the prevailing market prices for their crops.

The starkest instances of poor farmers' reliance on landowners were noticeable in lower Sindh. I travelled to a remote sharecropping village in Chor union council[7] of the Umerkot district in Sindh, consisting exclusively of around 100 sharecropping households, all of which belonged to the Bheel community.[8] Being landless, all these families were involved in sharecropping arrangements with about a dozen large landowners in adjoining villages. Three local landowners owned hundreds of acres each and employed several sharecroppers from this and other nearby villages. I had conversations with several poor farmers who worked on a cropsharing basis. They pointed out how their share of the cost of inputs came to around Rs 20,000 for cultivating an acre of land. Besides sharing the cost of fertilizers and pesticides, they had to singularly bear the cost of grain threshers – since local landowners refused to share this cost with them. When asked why they did not try to lease the land instead of sharecropping, the villagers stated that the local landowners did not want to lease them the land and, even if someone agreed to do so, they could not afford to pay between Rs 20,000 to Rs 30,000 rent for an acre per annum, and meet additional input costs on their own. In fact, most of them mentioned that they were indebted to large landowners already since they had to borrow money not only for inputs, but to meet other necessary expenditures, including expenses incurred because of weddings, funerals, and illness.[9] These sharecroppers had borrowed between Rs 30,000 and Rs 100,000 from different landowners. Some of the sharecropping families send family members to other districts in the province (Sukkhur, for instance) for seasonal labour, while other sharecropping households engage in seasonal labour on agricultural lands in their surrounding vicinity. Local landowners who did not have access to threshers had to hire seasonal workers.[10]

These sharecroppers admitted that their family members (women and children) had to help with the cultivation process, during the time of mustard or sunflower harvesting, for instance, but these additional family members did not receive any remuneration. Those indebted to the landowners also mentioned that their children were often called in to work in the landowner's homes for no remuneration except food. Most of the interviewed sharecroppers were selling their crops through the direct involvement of landowners, who in turn had connections with commission agents, or else would take the wheat to sell at government procurement centres. These sharecroppers were unaware of the price differentials between what they were paid and what price was settled on between their landlord and the commission agents, or even the procurement price set by the government during a particular year. Landlords would give sharecroppers their share of the crop in cash, subsequent to deducting interest and partial payments for loans extended during the previous year. The inability of poor farmers to question landowners about the outcomes of this informal accounts mechanism, along with the numerical illiteracy of poor farmers, made it difficult for them to determine the actual value of their labour.

Some of the sharecroppers I interviewed had the discretion to decide how much of their wheat they wanted to sell, whereas others had less say in this matter. A few of the surrounding landowners were taking the entire share of crops belonging to sharecroppers, and then releasing a few kilograms per month for household consumption until the sharecroppers' share of the produce ran out. The landowners portrayed this measure as being in the interest of sharecroppers, so as to ensure sharecropping households a steady supply of wheat till the next harvest came in, and to prevent them from accruing more loans. However, the sharecroppers complained that if they ran out of wheat during a given month, they would have to buy their own wheat at the same rate as flour was being sold in the market (which was a relatively higher price than the price they had been compelled to sell their wheat for at the time of harvest).

Sharecroppers were, however, not only being exploited by large landowners but also by intermediaries such as commission agents. In districts where sharecropping has been declining in favour of leasing, poor farmers could no longer rely on large landowners to purchase fertilizers or seeds. Large landowners were instead demanding a fixed advance rent, in lieu of the provision of land to sharecroppers. Poorer farmers thus had to rely on intermediaries such as commission agents for advance payments against the sale of their crops, in order to purchase required inputs such as pesticides and fertilizers. Some of the literature on globalization acknowledges the fact that middlemen margins limit the pass-through of the benefits of trade liberalization to primary producers, such as poor farmers (Mahmood *et al.* 2010, 2012). Commission agents themselves, however, insisted that they take their due share of margins in order to perform vital functions to link poor farmers to the marketplace.

Many commission agents whom I interviewed were based in district-level grain market situated in close proximity to pesticide and seed dealers. Some of pesticide and seed dealers sold agri-inputs to farmers on cash, while others would even provide the required inputs on the basis of receipts issued by commission agents; receipts which commission agents would reimburse at the end of the day from their own pockets, and later deduct from farmer reimbursements after their crops had been sold at harvest time. Most of these commission agents claimed that they did not charge interest rates on the loans they extended to farmers, but they did admit to purchasing crops from farmers at lower prices and selling them on for a profit of around Rs 40 per *maund* of wheat, in a district like Gujrat, for example. Moreover, commission agents said they had to incur the additional cost of hiring labourers to collect crops from the farm-gate and transporting these crops to the wholesale market or to flour mills, since poor farmers could not afford to pay these extra costs at harvest time when their finances were most depleted. Having built up a lending arrangement with a particular commission agent, poor farmers had to concede to the price offered for their crops, after the cost of advance payments was deducted, since not doing so risked disrupting their only available source of credit provision and support to ensure their crops reach the marketplace. It is thus a combination of indebtedness as well as poor farmers' continued dependence to get needed services, which leads them to acquiesce, and sell their produce for less than the market price.

The level of dependence on commission agents was seen to vary from area to area. In Gujrat district, for instance, commission agents were purchasing wheat at between Rs 30 to Rs 50 less than the government set control price of Rs 950. Whereas agriculture department officials in Umerkot informed me that commission agents in their area, on average, purchased grain for Rs 800 or Rs 880 from poor farmers to sell it on to the Food Department for Rs 950. The general paucity of poor farmers in Umerkot, the lack of other economic opportunities, and more remoteness from the grain market, seemed to enable commission agents in the area to extract an even greater margin of profit.

The inability of poor farmers to store wheat grown by them, their desperate need for cash, and their reliance on commission agents to transport their wheat to the gain markets, compels them to sell their wheat for a lower price. These observations not only validate the findings of the above-mentioned NHDR survey concerning the prevalence of asymmetrical markets in rural Pakistan (Hussain *et al.* 2003), they further reveal the implicit compulsions due to which poorer farmers experience market distortions, whereby they get paid less for their outputs in comparison to larger landowners.

Yet, while market distortions continue to plague poor farmers, donor attempts have not given due attention to addressing the underlying causes of distortions which disadvantage poor farmers in the marketplace. Instead they have placed greater emphasis on increasing the ease of entry of multinational corporations (MNCs) in agricultural markets to bolster the rural investment climate and increase agricultural productivity through technological solutions.

Lessons emerging from neighbouring countries like India have not dampened the support for MNC penetration into the agricultural sector in Pakistan. World Bank-supported economic reforms, and the opening of Indian agriculture to the global market over the past two decades, for instance, including the use of genetically modified crops, has increased costs, while reducing yields and profits for many farmers, to the point of severe financial and emotional distress. A growing number of poor farmers are becoming trapped in a cycle of debt and despondency. As a consequence, nearly 250,000 poor farmers in India have committed suicide between 1995 and 2009 (Center for Human Rights and Global Justice 2011).

Despite the fear expressed by local NGOs that similar trends will emerge in Pakistan, multinational corporations have managed to create an evident niche for themselves in Pakistan for the provision of agricultural inputs including seeds, fertilizers, and pesticides. Agri-business giants like Monsanto are poised to introduce genetically modified (GM) crops in the country.[11] The application of international patent rights for GM seeds is also obligatory for Pakistan, given that it became a signatory to the World Trade Organization (WTO) Agreement on Trade Related Intellectual Property Rights in 1995 (WTO 1995). Monsanto's claims have been further strengthened by the Plant Breeders'

Rights (PBR) Bill 2012, which acknowledges the patent rights of MNCs such as Monsanto to protect their right to profit from trade in GM seeds (IPO-Pakistan 2012).

A retired senior government official maintained that intellectual property rights also enable patenting and protecting local seed varieties and knowledge, to prevent MNCs from exploiting indigenous resources without adequate compensation or acknowledgment. However, the dearth of support mechanisms available to poor farmers to patent traditional seed varieties, or to take legal action against potential MNC infringement, has led many civil society organizations to show apprehension concerning the ability of the formulated PBR Bill 2012 to protect the interests of indigenous cultivators. Moreover, while ensuring patent rights, the PBR Bill 2012 does not contain provisions which would make MNCs responsible for hazards and damages to the environment, or human health, caused by genetically modified seed varieties, despite numerous emerging research studies pointing to such potential problems (Cellini *et al.* 2004; Séralini *et al.* 2007).

Besides decision-makers in Pakistan not paying sufficient heed to the above apprehensions, market-led approaches to agriculture, endorsed by overarching donor policy frameworks like the PRSP 1 and 2, continue endorsing the need for technological solutions to boost agricultural productivity. While the use of genetically modified crops is not mentioned categorically, the PRSP 2 in particular points to the need for protecting intellectual property rights,[12] facilitating technology transfer, and commercialization, as the means for 'Increasing Productivity and Value Addition in Agriculture' (Government of Pakistan 2008: 82).

Multinational involvement in the provision of pesticides and fertilizers is longstanding within the Pakistani context. Syngenta, a prominent Swiss MNC with a presence in 90 countries, for instance, has established it presence across rural areas of the country (Syngenta 2012). While it deals in seeds as well, it is in pesticides where Syngenta has a more prominent presence, controlling over 20 per cent of the total market share. While Syngenta launched its business operations in Pakistan in 1972, it has now created a more prominent role for itself by transcending the traditional route of supplying its products through traditional dealers, and launching its own franchise system called Naya Savera.[13] I came across a Naya Savera employee while I was walking around Kotpur village in the Gujrat district. I noticed a prominent lease tenant, who cultivates over 100 acres of land, getting a plot of his land sprayed by this Naya Savera employee. A Naya Savera field officer was also present there, and I asked him to tell me a bit more about their company and what he thought gave them a competitive edge over other local pesticide and seed companies working in the area. The Naya Sawera field officer claimed that unlike local pesticide or fertilizer dealers, their company charged fixed rates and did not cheat customers by raising prices arbitrarily. The quality of Naya Sawera products, and the results they produce in terms of boosting yields, were said to help sell their products even if they were priced higher than unregulated agricultural inputs, even if only well-to-do farmers could afford to purchase these inputs.

Interestingly, the Naya Sawera representatives also pointed out other supplemental activities being undertaken by the company to create a larger niche for itself in the local market. These included arranging community meetings across villages to create product awareness. This MNC (Sygenta/Naya Sawera) was using demonstration plots to showcase the benefits of its products to other farmers in the vicinity. Unlike the government's agricultural extension service, which just provides technical advice, this MNC was supplying new products free of cost to farmers who volunteered a piece of land for demonstration purposes. Such marketing schemes seemed to work readily, given that this fixed-rent tenant had provided a two *kanal* plot of land within the same village, for trial of a new pesticide for wheat crops.

However, this increasing prominence of MNCs, for not only providing inputs, but also advising farmers about how and what to use to grow better crops, was traditionally the domain of agricultural extension services. MNC encroachment into the realm of providing farmers technical advice has been allowed in developing countries like Pakistan based on the rationale that public extension services are outdated and inefficient. Instead of supporting extension services to overcome their deficiencies however, and to reach out more effectively to poor farmers, donor agencies have emphasized the need to either sub-let some of the functions of public extension services to private companies or else to dismantle the public extension system altogether and replace it with one or more specialized private service providers (Davidson *et al.* 2001).

Entities like the World Bank understandably favour such interventions as they further the goal of liberalization, and help curb public spending. To some extent this measure seems justified, given that the inadequacy of extension support services for poor farmers was, in fact, evident in several villages, with officials lacking resources to reach out to farmers. However, whether MNCs (or even local private companies) can step in to fill this gap and cater to the needs of poor farmers is not likely. As of yet, MNCs like Sygenta/Naya Sawera are replacing, instead of supplementing, government efforts to improve the efficiency and outreach of extension services. There was also no evidence of private sector companies like Syngengta/Naya Sawera trying to collaborate in any manner with the government's Agricultural Extension Department, even though they were found to be performing functions meant to be undertaken by agricultural extension workers. While they may be more efficient, profit-driven MNCs aim to focus on more progressive farmers who can afford to purchase their products. There is also an implicit conflict of interest in this situation, since commercially motivated companies are advising farmers concerning which, and how much of, particular agri-inputs are needed for maximizing yield.

Thus, donor-based endorsement of an increasing role for the private sector in the agricultural sector, instead of creating a 'level playing field for all', seems to yield rather unintended results in rural areas of Pakistan. Poor farmers in Punjab and Sindh seem to be in danger of getting caught up further in an exploitative cycle of debt due to the increasing reliance on the profit-driven private sector for agricultural development. They have to rely on intermediaries (such as landowners and commission agents) to continue participating in the increasingly commercialized agricultural production system, which is further exacerbating their exploitation. Nonetheless, the use of market mechanisms within the agricultural sector keeps growing. Besides encouraging the private sector to provide agricultural inputs to increase productivity, the market mechanism is being applied to creating more robust land markets, a precondition for which is computerizing the manual land management system in Pakistan.

Will computerization of land records help poor farmers?

The need for the computerization of land records is being justified by donors like the World Bank as the means to help resolve a constant source of conflict and expensive litigation, which is particularly troubling for poor farmers (World Bank 2006). This sub-section takes a closer look at an ongoing programme attempt to computerize land records in the province of Punjab, in order to ascertain its potential benefits for poor farmers.

There are approximately 190 million land records, purportedly containing the details of approximately 50 million landowners, in Pakistan. These land-holdings are also spread over a massive area. Punjab alone has a total area of 205,345 square kilometres (Revenue Department 2012). Manual land records provide inadequate proof of ownership, and they are not linked to spatial data to accurately identify specific plots. The antiquated manual system, coupled with corrupt practices of low-ranking and underpaid officials (*patwaris*), creates frequent inaccuracies in record keeping, leading to disputes, and also hinders the transfer and sale of agricultural property (Qazi 2005).

The increasing pressure of maintaining an increasing number of land records, due to intergenerational subdivisions of landholdings, adds further pressure on revenue records maintenance. In the Punjab, for example, around 8,000 *patwaris* are tasked with keeping land records pertaining to some 20 million landowners (Revenue Department 2012). Getting a *patwari* to discharge his official duties is difficult and often expensive. Many farming communities complained that it was hard enough for middle-sized farmers to get *patwaris* to do their work, let alone poorer farmers or women. Given the burden of work allocated to a *patwari*, it is also understandable that he is often genuinely busy and has to make people wait for long periods of time or make repeated visits in order to fulfil the cumbersome requirements of his job. I was however informed by less affluent farmers that a *patwari* usually gives priority to doing work for landowners who can afford to pay him some money on the side, or can do him other favours.

The very process whereby land mutations[14] take place is also murky. For instance, I witnessed this procedure first-hand in the Gujrat district, at the office of the *gardawar*[15] for the Mangowal union council.[16] When I arrived for my meeting with this *gardawar* at a dilapidated *haveli*[17] belonging to an old local landowning family, which served as his office, I found several local villagers assembled around him in the open veranda, including the *numberdar*[18] of one of the nearby villages. Noticing that I had arrived for our scheduled meeting, the *gardawar* told me to be seated nearby, since he was in the middle of a discussion.

I soon gathered that the discussion concerned an inheritance case, and the villagers had gathered so that the latest information concerning the transfer of landownership could be recorded in the mutation register. While I was unable to interject to get clarifications concerning the exact location of the village or the particulars of this case, it was evident from the ongoing discussion that the practice of altering the mutations register could be rather discretionary in practice. Before legalizing the land record, the *gardawar* asked the *numberdar* to verify the exact details of the inheritance case, including the exact names of the children of the deceased, except the ones who were accompanying him. The *numberdar* began to fumble and could not name all of them. This was a potential problem since inaccurate documentation would require a major subsequent effort by legitimate inheritors to have the land records corrected. It was also possible that some of the children of the deceased person were not being named, in order to deny them their land rights. However, the *patwari* intervened at this stage to inform the *gardawar* that the *numberdar* was unfamiliar with the details of the inheritance case since he had been away from the village for the past few years. The *patwari* further asserted that that everyone else present knew the exact details of the case as they were all directly related to the deceased. Upon these assurances, the *gardawar* proceeded with the formalities of altering the landownership record.

He then asked the *numberdar* to provide his national identification card. It turned out that the *numberdar* was not carrying an identity card either. The *gardawar* grew more frustrated and told all of the villagers to go away and bring back the required documents if they wanted him to alter the land record, or else to take up the matter with the *tehsildar*.[19] After listening to this, the villagers moved to another room, along with the *numberdar* and *patwari*, allowing me to proceed with my scheduled interview, perhaps waiting to renew their request once the *gardawar*'s temper had cooled down.

While I did not follow up further on the outcome of the above case, the above-described scenario itself illustrates the ample possibilities for inaccuracies to occur in the land records. The *gardawar* himself did not go to the village to verify how many children the deceased had left behind, and was relying on the word of only those village members who had come to his office, to establish the veracity of the inheritance claims. It is thus not hard to imagine how a rightful inheritor, either a son who was away or else some of his daughters, could be left out from identification as legitimate inheritors of the deceased elder's property. Such inaccuracies within land records do, in fact, occur regularly, exacerbating land-related disputes.

Numerous other legal disputes can also occur around the issue of agricultural land rentals, including accusations of illegal possession, undue eviction of tenants, as well as problems related to the recovery of rent. Varied cases of land disputes are heard in Revenue Courts as well as in Civil Courts, but the lack of credible records being readily available causes major delays in resolving legal cases (Qazi 2005). Donor agencies like the World Bank thus point to the inefficiency and unreliability of the land record management to be causing high transaction costs, repeated disputes concerning ownership and tenancy rights, as well as becoming a barrier to entering lease arrangements which in turn discourages greater investment in agriculture (World Bank 2007b).

The manual land administration system has also been criticized for having a particularly adverse affect on poor farmers, who lack resources and leverage to grapple with cumbersome bureaucratic processes, making them vulnerable to the predatory behaviour of corrupt officials such as *patwaris* (Qazi 2005; World Bank 2007b). Strengthening land titling is therefore considered to be especially beneficial for poorer farmers, based on claims that 'clear land rights have far-reaching implications for social cohesion and governance acting as an important catalyst in stabilizing communities, empowering individuals and reducing social exclusion' (World Bank 2006: 1–2).

It was on the basis of this rationale, that the World Bank began funding a major Land Record Management and Information Systems (LRMIS) Programme in the Punjab in 2006. It is important to note, however, that the most immediate emphasis of the LRMIS programme is on computerization of land records. This implies the computerization of existing land records, in consultation with district government officials, rather than undertaking fresh surveys to make land records more accurate. The LRMIS had completed land record computerization in seven out of the 36 districts of Punjab by mid-2012 (Revenue Department 2012), and work on computerizing land records in the remaining province was also ongoing.

I asked one of the most senior officials in this LRIMS programme, headquartered in Lahore, how the computerization of existing land records would help poor farmers, especially landless sharecroppers. I added the caveat of several existing land records not mentioning poor sharecroppers as cultivators, which is a typical way in which *patwaris* serve to protect landowners from being legally compelled to respect the rights of sharecroppers under the existing tenancy laws. Initially, he seemed a bit nonplussed by the question and did not answer it immediately. Later on during the interview, when we returned to the problem of inaccuracies within existing land records, he claimed that such discrepancies would be addressed in time due to the ongoing process of computerizing land records. One of the suggested ways in which this could happen relied on *patwaris* providing this information to LRMIS, the next time they go to villages within their jurisdiction to undertake crop inspections for revenue collection purposes, which is the time *patwaris* are meant to note down who is actually cultivating the land. When asked how the continued reliance on *patwaris* to obtain such information would make any difference, since it was *patwaris* themselves who manipulate land records in favour of landowners by avoiding documentation of tenants working on their land, he did not have a response. The other suggestion put forward by the LRMIS official did not involve *patwaris*, but was based on the assumption that villagers could themselves come and check land records at LRMIS service centres across the province, for a small fee, and put in an application to alter any inaccuracies. Expecting illiterate or semi-literate poor, and often indebted farmers, to go and check if their names have been documented as sharecroppers on computerized land records, and then challenging the veracity of these records, if inaccuracies were to be found, seemed rather optimistic.

Some of the other stakeholders I spoke to were also not very optimistic about the LRMIS programme's potential to help poor farmers. For example, an expatriate project manager for a Department for International Development (DFID) funded project,[20] working on a public sector reforms project with the Punjab government, recalled numerous problems with a similar initiative in Andra Pradesh in India back in 2002. The computerization programme in Andra Pradesh had also failed to achieve similarly stated pro-poor results. An evaluation of the Andra Pradesh computerization effort found it to have succeeded in automating some of its manual processes. Yet, these successes were limited to 'back office' operations, and thus citizens did not experience a real degree of simplification of their experience with the registration process. Poorer citizens in particular were found to be denied access even to the interior of the office by the document writers and other middlemen, and it thus remained difficult for them to access the computerized land record system. Most procedures and requirements remained incomprehensible, which were also not found to be available to them in their vernacular languages. Such problems became a major hurdle preventing ordinary citizens from benefiting from the land record computerization process (Centre for Good Governance 2002). Poor farmers could similarly be subjected to discrimination within the LMRIS programme, with LRMIS service centres being as, or more, inaccessible to them as a *patwari*, who at least comes to their village.

The computerization of land records, and the resulting possibility of generating expeditious tenancy agreements, may conversely encourage larger landowners to undertake more tenancy arrangements with farmers, or even agri-businesses, instead of relying on poor farmers to cultivate their surplus land on a cropsharing basis. Computerization's potential for encouraging increased leasing arrangements may thus provoke further evictions of sharecroppers, and lessen land access for poor farmers, who cannot afford to cultivate land by paying an upfront lease amount. Similar trends became visible when large landowners were subsidized in their use of mechanization, which lessened their reliance on the agricultural labour force during the 'Green Revolution' in the late 1960s and 1970s (Alavi 1973). Alterations in land use can cause major upheavals, as was seen recently in the case of a widespread resistance movement trying to convert sharecropping arrangements into lease contracts on military farms in the Punjab (discussed in detail in Chapter 6).

Lessons emerging from the 'Bhoomi program' of computerizing land records in districts surrounding Bangalore in the Indian state of Karnataka are also instructive. This project aimed to digitize 20 million land records, and is being hailed by the World Bank as a 'best practice' model of 'e-governance' to be replicated in other parts of India, and elsewhere in poor countries (Chawla 2004). However, ethnographic research on this initiative indicates that the computerization of land records has led to increased corruption, and further inaccessibility of records for poor farmers due to problems including illiteracy and the need to travel to offices located further away from villages. Conversely, this project was seen to be facilitating very large players in the land market to capture vast quantities of land (Benjamin *et al.* 2007).

Computerizing existing land records may facilitate expedient land transactions in Pakistan as well, but doing so could easily enable large landowners to lease lands to capital-intensive agri-businesses, which does not offer significant employment opportunities to poor farmers and agricultural labourers. Besides facilitating more accurate revenue generation, transparent land records could potentially make land redistribution more effective, yet this possibility is nowhere evident in the rationale or programmatic objectives of the World Bank funded LRMIS programme.

The LRMIS programme also lacks a gender component. A range of factors, including land tenure laws, customs, social norms, and social relations, combine to continue excluding women from the ownership and control of property (especially immovable property like arable land) across South Asia (Agarwal 1995). The World Bank reiterates the need for special attention to cultural and economic constraints faced by rural women (World Bank 2007b), yet the World Bank-supported LRMIS programme in Punjab has made no explicit attempt to address these problems.

The LRMIS programme, with its evident focus on computerizing land records, cannot address the inaccessibility and complexity of the legal requirements and bureaucratic processes which diminish the chances of women claiming and acquiring their existing land rights. Nor does LRMIS offer the potential to side step the patriarchal values due to which women forfeit their existing rights to landownership. Reflecting on this problem, a senior official in the Law and Parliamentary Affairs Department for the Government of Punjab in Lahore offered an interesting suggestion. He mentioned how donor pressure had compelled the government to employ female officers in a range of other departments, including the police force, and he questioned why the World Bank could not have made a provision for female *patwaris*, to ensure that women are not deprived of their land rights. There is no evidence of the World Bank doing this in Pakistan.

The LRMIS also has not taken into account other social constraints, like the lack of women's mobility, despite lessons emerging from other places where the computerization of land records has taken place. The Bhoomi project in Karnataka particularly found it difficult to for women to obtain land records despite computerization of the earlier manual land record system (Chawla

2004). While most poor and illiterate farmers were struggling with computerization, women in particular were seen to be at a loss comprehending the
new system. Women in Karnataka used to own 12 per cent of the land prior
to computerization, but there was no gender-disaggregated data available to
determine how land record computerization had changed this situation.

Whether focusing on ensuring secure land tenures or facilitating land
transfers are adequate measures to alleviate rural poverty is debatable. Such
attempts do not contend with the political economy of landownership, due to
which existing land records are inaccurate and often biased in favour of large
landowners. The computerization of land titles has been criticized for its
rather technocratic approach, placing emphasis on administrative notions
such as the need for improving 'transparency' and 'efficiency' (Benjamin *et al.*
2007), while remaining oblivious to underlying imperatives which can subvert
these goals, especially the power relations associated with landownership.

Using intermediation to bring poor farmers into rural land and capital markets

Donors like the World Bank have been using NGOs to intervene on behalf of
the poor in capital markets through the provision of micro-finance for the
past two decades. More recently, the same logic of using NGO intermediation
in capital markets has been used to involve NGOs to intervene on behalf of poor
farmers in land markets, in order to ensure their access to land. This section
considers both these types of interventions and assesses their potential relative
to improvements in the lives of poor farmers.

The need for credit plays an important role in agriculture, especially since
the 'Green Revolution' in the 1970s, which led to an increasing reliance on the
use of expensive agricultural inputs such as fertilizers and pesticides. Initially,
institutional credit for agriculture was available to farmers in the form of
taccavi loans,[21] before cooperatives and commercial banks became increasingly
involved in the provision of agricultural sector lending (Government of Punjab
1958). Delays and procedural difficulties in the sanctioning and disbursement
of loans rendered the system of *taccavi* lending inefficient, and ultimately
these loans were discontinued in the mid 1990s. Cooperative lending was
encouraged as another source of credit for agriculture. However, despite the
development of over 60,000 cooperative societies, the misuse of subsidized
funds, and the incapacity of the government to effectively monitor and regulate the activities of cooperative societies, led to their failure. Two specialized
institutions, namely the Agricultural Development Finance Corporation and
the Agricultural Bank, were also established, and then merged to form the
Agricultural Development Bank of Pakistan (ADBP) in 1961. The ADBP has
been used as a subsidy delivery vehicle by successive governments, but it often
lent large sums to prominent landlords, who subsequently defaulted on their
loans and had them written off using their political influence. Subsidy-driven,
and often irrecoverable, agricultural credit programmes have largely proven

ineffective in alleviating rural poverty, or spurring agricultural production, and they have mostly excluded poor farmers from accessing available credit lines (Hussein 2009).

Given the huge, unmet, demand for agricultural credit in Pakistan, commercial banks have tried to develop specific lending instruments for the agricultural sector, for instance, to enable the purchase of tractors, or to undertake land development. Yet, the manager of the Mangowal branch of a prominent commercial bank in the Gujrat, admitted that the internal requirement for commercial bank lending to farmers is predicated on the need for collateral. In the case of the bank where the respondent worked, the lending requirement for agricultural loans was a minimum ownership of 5 acres of land, meant to serve as collateral. However, this bank manager argued that his institution was a more efficient lender than the government's Agricultural Development Bank, since there was no political or bureaucratic influence over its lending operations. Being a commercial entity, he argued, further compelled his bank to exercise greater vigilance to prevent defaults in comparison to the government subsidized ADBP. He did, however, complain about a current crunch on all banks, due to heavy government domestic borrowing, which has left little money for other borrowers, even for landowners who meet the criteria of availing collateralized loans.

Because of their lack of access to formal credit, poor farmers have generally relied on informal sources for meeting most of their increasingly expensive agricultural input cost requirements. Informal providers serve a predominantly lower income group, who are perceived by the financial institutions as 'un-bankable' due to their inability to comply with conventional loan collateral requirements. Although it is difficult to properly assess the actual scope of this informal lending, the Agriculture Census of 2000 estimated that 65 per cent of the outstanding debt of all rural households was being provided by non-institutional or informal sources (Hussein 2009). This informal credit market is served by a wide variety of providers, including, predominantly, family and friends, landlords, moneylenders, traders, and commission agents.

While the problem of large farmers usurping available credit facilities for agricultural development is well recognized (Haq and Khalid 2011; Hussain 2008, 2009), it was interesting to observe how some informal lenders had actually been borrowing credit from formal sources, even commercial banks, and passing on this credit to poor farmers. The branch manager of the above-mentioned commercial bank in Gujrat, for instance, pointed out that while interest rates for agricultural loans have begun rising, they were still very low (as low as 8 per cent). These low interest rates did not enable poorer farmers without required collateral to avail the credit facilities, but it did make it easy for commission agents to borrow money from such banks, and lend it onwards for a slightly more inflated interest rate to poor farmers. This finding was confirmed by commission agents in the local grain market, who said that they have had to put their businesses, and even private property, as collateral to secure loans, so as to make advance payments to poor farmers for investing in their crops. The commission agents, however, did not see this onward lending as being exploitative, since they did not admit that they charged interest from the poor farmers to whom they loaned. They did, however, acknowledge that their provision of credit to poor farmers compelled these farmers to sell their produce to the commission agents below market prices. Commission agents considered this practice justified, however, given that they put themselves under debt to provide poor farmers the credit to purchase necessary inputs, and they also provide them the service of picking up crops from their farm-gate, for which there is no other institutional support.

Donor agencies like the World Bank do recognize the lack of access to credit for poor farmers, which is considered a major reason for lingering rural poverty (World Bank 2007b). This perception has led to an emphasis on the provision of micro-finance services, including micro-credit loans to the rural poor.[22] The Aga Khan Rural Support Programme (AKRSP) in the Northern Areas, and the Orangi Pilot Programme in urban Karachi, were among the first NGOs to initiate micro-finance programmes in Pakistan. These programmes helped spawn other rural development initiatives across the country, such as the National Rural Support Programme, the Punjab Rural Support Programme, and the Sindh Rural Support Programme. These were all patterned along the lines of AKRSP, and funded by the Livestock, Health and the Environment ministries, multilateral agencies like the Asian Development Bank, and bilateral donors including United States Agency for International Development, and the United Kingdom's Department for International Development (Rural Support Network Programme (RSPN) 2012).

Encouraged by donor funding, many other NGOs also began micro-finance programmes in the 1990s. At the government level, concerted efforts for the promotion of micro-finance began in the year 2000, when an apex funding body, the Pakistan Poverty Alleviation Fund (PPAF), was established for poverty alleviation. As an apex organization, the role of PPAF is to act as a wholesaler and intermediary of funds with donor support. In 2000, the government, under a separate ordinance and with financial assistance from the Asian Development Bank, established the Khushali Bank for providing micro-finance to the rural poor in un-served and under-served areas of the country (Hussein 2009).

The micro-finance sector now consists of a diverse range of players. It includes development NGOs, micro-finance and development finance banks with micro-finance operations, and commercial banks offering a range of micro-credit and micro-finance products. Yet most micro-finance providers in Pakistan are unsustainable. They depend on donor funds and subsidized lines of credit from the PPAF. The micro-finance sector in Pakistan grew exponentially with client outreach jumping from 241,016 borrowers in 2003 to almost 1.5 million borrowers in 2007. Average annual expansion stood at approximately 61 per cent during this period. This growth trend has slowed down since then, with the sector adding barely 600,000 borrowers between 2007 and 2010 (Haq and Khalid 2011: 1). Questions are thus being asked about the effectiveness of micro-finance lending to extend sufficient outreach to alleviate poverty across the country.

Due to the high cost of credit delivery to the doorsteps of the poor, the accompanying capacity building programmes to ensure the productivity of their clients, as well as the need to ensure self-sustainability, micro-finance institutions charge much higher interest rates than commercial banks in Pakistan. Their impact on poverty alleviation resultantly remains controversial. Many of these entities are being criticized for being extractive and overly commercial, instead of trying to lift people out of poverty (Bateman 2010).

During my research in a village in the Badin district in the province of Sindh, I discovered poor farmers taking money from lenders to pay off micro-credit entities, and vice versa. I went to one such village located some 40 km away from the Badin district headquarters. According to a village profile prepared by NGO personnel involved in provision of micro-credit in the area, sugar cane, rice, and wheat were the main crops being grown in this given village, and the villagers also reared 56 cows and 81 buffaloes to supplement their household incomes. There were only 28 households in the village and a total population of 126 residents. There were only 65 acres of cultivable land and 15 acres of uncultivable land in the village. Out of the 28 households, the largest landowner in the village had 12 acres of land, while some other farmers had smaller parcels, with the minimum ownership being 1 acre. A total of 15 landless farmers from the village were working for a landlord in the adjoining village who owned 275 acres and had 25 sharecroppers working for him. Those working on a 50:50 cropsharing arrangements had to share half the input costs, such as tractor rentals, whereas those who asked the landowners to pay for these expenditures had to work on a 25:75 cropsharing basis (with the landowners took 75 per cent of the share of the crop). Another four people from this village had also given this large landowner their land on lease, since they themselves did not have the resources to cultivate it, but the amount of rent they accrued from the leased land was meagre; only Rs 4,000 per annum (much less than lease rates in districts like Gujrat, where the annual lease rate was at least four times this amount). The lease rates were low in this village since the quality of soil was poor in the region and water scarcity required the use of expensive diesel run pumps to draw up ground water to irrigate crops.

The NGO with which I visited the village had classified it as a poor locality, and had extended micro-finance to five borrowers in addition to providing micro-insurance services to 11 (7 men and 4 women) village residents. I spoke to three out of the five micro-finance borrowers in the village. It was surprising to note that all of the three borrowers were also taking loans from *baniyas* (Hindu money lenders who have been providing informal credit in the area since before independence in 1947).

One of these micro-credit borrowers was an Intermediate (Grade 12) graduate, and had taken five cycles of NGO-provided micro-loans. He had started by borrowing Rs 5,000 and had qualified to borrow Rs 20,000. He had set up a small shop with this loan, in addition to using loaned money to pay for inputs for cultivating 2 acres of land himself, and for sharecropping another 2 acres (0.81 hectares) with his brother. This borrower was taking money simultaneously from the *baniya* the entire time that he had been a micro-credit borrower. At the time of the interview, he had borrowed a Rs 20,000 loan from the NGO, and Rs 14,000 from the *baniya*. While the preceding year he took Rs 8,000 from the *baniya*, he had to borrow more from him during the current year (when fieldwork was conducted) to keep the dual interest repayment cycles going. This borrower admitted he found it very difficult to keep up with paying back the high interest-rate loans of the *baniya* and the NGO, and also admitted that he often took one loan to help pay off the other. During a past year he had to sell two goats, and another year he sold a calf, to service his loan repayments.

Despite the above scenario, the Micro-Finance Officer, and another senior officer in charge of this large NGO's field office in Badin, justified their lending programme. They claimed that poor famers lack access to official credit facilities such as those offered by the Agricultural Development Bank, which are primary usurped by large landowners. Hence, poor farmers without access to micro-finance are compelled to borrow money from the informal sector, where lenders not only charge high interest rates, but also pass on debt burdens to the next generation, in case the borrower dies.

Based on assertions that micro-loans ensure poverty alleviation, by providing poor people an alternative form of credit which does not require collateral, nor subjects them to exorbitantly high interest rates charged by informal sector money lenders, prominent NGOs in Pakistan have secured support from major donor agencies to spread their operations across the country, not only in Sindh. While generalizations cannot be drawn from the above example, it can be argued that in cases where lenders are using micro-finance to pay off other informal sector lenders (such as *baniyas*), and/or vice versa, micro-finance is not fulfilling its stated goals. When micro-finance organizations lend to groups, they can pass the risk of individual clients defaulting onto the entire group. From the institution's perspective this makes sense, since default can be avoided if the rest of the group repays the loan. But from the clients' perspective, one person's default implies hardships for everyone belonging to a peer group. When other group members are forced to make up the difference, the debtor also faces censure, and sometimes even violence, from the other members (Hulme 2003). The micro-lending operations described above can be defined as a predatory form of lending, when they do not take into account the ability of borrowers when issuing loans (Cecelia 2008).

The compulsion of micro-finance entities is to extend their lending outreach wide enough for their operations to become financially sustainable. In order to do so, they need not only get more clients, but also charge high interest rates from poor borrowers, who have to pay high interest rates to be integrated into the market economy (Hulme 2003). While most MFIs in Pakistan have not yet reached financial sustainability, and remain dependent on donor funds, they are also guided by the same imperative to extend their client out-reach. The high interest rates charged by these entities can often exclude the poorest of the poor, if they lack the capacity to become productive enough to pay back pre-scheduled interest payments as soon as the micro-lending cycle commences. Although micro-finance entities aim to provide loans to bor-rowers who have the capacity to repay their loans, the compulsion to broaden their client base has led to an evident trend of lending to already indebted borrowers. In certain instances, as in the above-mentioned cases of villages in the Badin district, mainstream micro-finance entities have been compounding pressure on the poor, further exacerbating the vicious cycle of indebtedness. This phenomenon, of taking micro-finance loans to pay off other lenders, and vice versa, could lead to more severe indebtedness. Nonetheless, micro-finance is still considered a relevant tool by donor agencies like the World Bank for helping overcome rural poverty in Pakistan, and it is also being used to ensure the sustainability of state land distribution beneficiaries under the Benazir Landless *Hari* Scheme (as discussed in Chapter 3).

Despite the fact that the impact of micro-finance on poverty and empower-ment is not a foregone conclusion, donor agencies, including the World Bank itself, keep reiterating the utility of micro-finance and the use of NGOs to provide poor people access to credit. Some critics of micro-finance have even termed such interventions to be part of a neo-liberal conspiracy, whereby the poor, at best, can scrape a precarious living through micro-enterprises, and are thus prevented from uniting to improve their position (Bateman 2010).

Predicated on the same principle of devising market-based approaches to addressing the plight of poor farmers, I interviewed development consultants working under auspices of a large bilateral donor agency-funded development programme being implemented in the Punjab, which has formulated a pilot programme aiming to improve access to land with the assistance of NGOs. The pilot programme proposed using NGOs to lease large parcels of land from large landowners and then sub-letting this land onto poor farmers on a cost recovery basis in southern Punjab. This intervention aimed to address the declining trend in sharecropping in favour of lease arrangements, which is further aggravating the problem of poor rural households' access to land as they are unable to pay upfront rents to then cultivate the land. The proposed project was thus designed to use NGOs to organize landless and marginal farmers into community groups and, in turn, intermediate on behalf of these groups in land-rental, tenancy, and credit markets, by providing them sup-plemental mechanisms for risk mitigation and access to working capital in the form of micro-finance lending. The intervention was formulated in a manner which could avoid the need for purchasing land for landless and marginal farmers, by instead aiming to increase land access among the landless and marginal farmers by enabling them to enter the land tenancy and land rental markets.

Lack of resources led this intervention to be delayed, so it was not possible to assess its actual impact on the poor. Yet the very design of the project is indicative of another market-led intervention to contend with the persisting problem of the inequality of landownership without resorting to land redistribution, or contending with the range of underlying power dynamics, which distort market mechanisms.

Corporate farming and poor farmers

Donor agencies like the World Bank have been emphasizing the need for more investment in the agricultural sector in developing countries generally, as well as in Pakistan (World Bank 2007a, 2007b). This donor endorsement has encouraged foreign equity firms to bring in capital investment for high growth agri-business projects (Tirmizi 2012). Besides offering profit making opportunities to private enterprise, such trends have also been described as evidence of an increasingly commercialized global food production system, within which seemingly opposing phenomenon of agricultural surpluses and growing food insecurity can easily co-exist (Barbier 2000; Kugelman and Hathaway 2010). Yet, the World Bank-endorsed Poverty Reduction Strategy Paper for Pakistan has also categorically stressed the need for corporate farming to improve agricultural productivity and profitability through high-tech interventions, which in turn are expected to produce 'high value products, value addition at the farm level, provide job opportunities and exportable surplus, and ensure international competitiveness' (Government of Pakistan 2003: 47).

Yet, the assumed benefits of corporate farming, especially in terms of it being deemed a vital component of a strategic document to address poverty are disputable. Earlier sections of this book have already referred to the controversy unleashed by the government decision to lease out state-owned land under the Corporate Farming Ordinance 2001 to foreign companies to export food produced in Pakistan. An attempt is made now to draw attention to the practice of corporate farming as adopted by an influential Pakistani landowner.

Based on fieldwork in late 2011, conducted at one of the most prominent and profitable corporate farms in Pakistan today, the following subsections aim to particularly highlight how the Tarakee Farm (not real name), despite being a profitable and innovative venture, does not have a positive impact on poor farmers. Instead, by leasing vast amounts of land for its operation, Tarakee Farm has exacerbated the problem of land access for poor farmers in the surrounding area. Moreover, since the corporate farm relies on capital-intensive technologies, it offers limited work opportunities, with modest remuneration, mostly to female agricultural labourers and to male seasonal workers, who comprise the bulk of the workforce at this corporate farm. Despite its limited potential to aid poor farmers, the Tarakee Farm has, however, itself managed to secure donor, and even NGO, support, largely due to the owner of Tarakee Farm, who is an affluent landowner, industrialist, and politician, and has also held prominent positions in organizations working in the development sector.

a. The Tarakee Farm, its owner, and how it works

The Tarakee Farm is located in southern Punjab, within one of the poorest districts in the entire province.[23] However, it became the site for putting up one of the largest corporate farms in the country, owned by one of the most successful men in the country. The owner of the Tarakee Farm can be categorized as an ideal 'progressive farmer', a term which entities like the World Bank use to describe those farmers who are able to apply required technological and management practices to achieve high growth and, they are thus considered as 'good benchmarks' of what can be done to increase the envisioned agricultural productivity increase within the country (Ahmed and Gautam 2013).

The Tarakee Farm is cultivating 3,000 acres of land, growing a range of crops, but with an emphasis on cotton and mango farming. A large workforce is employed by the Tarakee Farm, comprising of 400 permanent employees, anywhere between 200 to 400 contract workers and up to 1,000 daily wage labourers.[24]

There is a lot of impressive infrastructure on the farm. This includes laboratories for bio-testing crop viruses, a fruit packaging facility, and a ginning mill, as well as water reservoirs where water for irrigation is stored. There are several office buildings, spacious residential houses for senior staff members, as well as recreational facilities including a gymnasium. There is also a hostel for junior ranking employees, as well as separate quarters for labourers and other menial staff, employed by the corporate farm on a permanent basis.

The Tarakee Farm owner takes an active interest in the management of the farm, in coordination with a team of professional managers, but he has allocated only a limited amount of time to this venture due to his numerous other commitments. To save time, he flies in on his personal jet, for which an airfield has been constructed on the farm. The Tarakee Farm owner is involved in another corporate farming venture in the nearby district of southern Punjab, in partnership with his brother-in-law, a very powerful man in his own right.[25] Their other corporate farming venture also involves growing sugar cane on 25,000 acres, for four sugar mills. Moreover, the Tarakee Farm owner is also an industrialist, with major shares in an international beverage company, and in paper-pulp, sugar, and textile mills. He was twice elected a parliamentarian, headed a provincial government taskforce on agricultural development in the 1990s, and served as a federal minister in the early 2000s. He is also actively involved in mainstream development interventions across the country, as a board member for the RSPs, and the World Bank-supported Pakistan Poverty Alleviation Fund, all of which are undertaking development projects in rural areas using market-based strategies such as the provision of micro-finance and other cost-sharing development initiatives.[26]

This background information, concerning the owner of the Tarakee Farm, has been mentioned here to not only illustrate the concentrated nature of power and affluence in Pakistan, but to also foreshadow the way in which the Tarakee Farm itself operates. The following subsections now begin focusing on various aspects of the Tarakee Farm, primarily with reference to their implied implications on the lives of poor farmers.

b. How Tarakee Farm is exacerbating inaccessibility to land

Significant alterations have been taking place in rural land use patterns, which have serious implications for poor farmers who already lack access to land, as is pointed out in Chapter 3. The growing decline in sharecropping arrangements noted in the literature is in contrast to increasing trends of self-cultivation and a greater emphasis on leasing land rather than giving it to sharecroppers for cultivation (DFID 2009). Moreover, my research on the Tarakee Farm found that the promotion of agri-business is another contributing factor impelling this changing pattern of land use, which also serves to compound land scarcity for poor sharecroppers.

A closer look in the preceding section, at how Tarakee Farm acquired the land for its operations, clearly indicates its preference for leasing land, which not only began driving up lease rates beyond the reach of poor farmers, but simultaneously decreased the amount of land available for sharecropping by poor farmers.

The area in which the Tarakee Farm is located used to belong to the Forestry Department before, and even for some years after, independence in 1947. Field Marshal (R.) Ayub Khan[27] began allotting this forest land to compensate people displaced by the transplanting of Pakistan's capital from Karachi to Islamabad in 1959, and subsequently those displaced because of dam construction in Mangla and Tarbela Dam over the next decade. These government allocations remained relatively small (under 100 acres), perhaps because Field Marshal (R.) Ayub Khan had also initiated the first redistributive land reforms attempt in 1959. The remaining forestry department land (around 1,400 acres) was taken over by the army. Retired army personnel, who were in turn provided this land, began selling it commercially.

The Tarakee Farm owner had bought 350 acres in this area in the mid 1980s, when he first decided to begin farming himself. He then began buying more land, and currently owns 1,700 acres. Back in 1995, Tarakee Farm owner was only farming on the land he personally owned, but as profits began increasing, he decided to begin leasing land as well. The Tarakee Farm was cultivating 13,000 acres of leased land, in addition to the land belonging to the owner and his family.

Several large landlords were still present in the vicinity of the Tarakee Farm, who owned a few hundred acres each. However, while these large land-owners used to take on poor farmers to work on their land under sharecropping, this trend steadily began declining. One Tarakee Farm manager estimated that the proportion of the local population engaged in sharecropping arrangements had declined from nearly 80 per cent to around 20 per cent over the past decade or two. This in turn had made it difficult for poorer farmers to continue farming, since they could not afford to pay upfront rents to acquire land for cultivation. On the other hand, senior management at the Farm admitted that they are able to rent land below the market rate, since they only deal with a handful (around five) of large landowners in the area, and sign lease periods ranging between three to five years. The Tarakee Farm's average leasing rate was around Rs 15,000 to Rs 16,000 per acre per month, well below the average market rate of Rs 30,000 to Rs 40,000 per acre per month. While the Tarakee Farm was able to rent land at reasonable rates using economies of scale, the farm's immense demand for leasing lands had increased land scarcity, driving up lease rents for poorer farmers who cannot afford to pay advance payments for lease arrangements over multiple years.

The Tarakee Farm's immense land requirement had exacerbated land scarcity for smaller farmers in surrounding villages. Poor farmers, who no longer have access to land on sharecropping due to this increasing trend of leasing land instead, were keen to engage in sharecropping with the Tarakee Farm. However, according to the farm's management, this was not a feasible option since ensuring that sharecroppers meet the high production targets of the Tarakee Farm would require too much supervision, and also require inputs and technical knowledge which poor farmers did not possess. The existence of these on-ground realities vindicated broader critiques of corporate farming, whereby large capital-intensive agri-business operations are criticized for displacement of poor farmers (Singh 2006).

Agriculture in Pakistan already experienced such displacement trends, when the government subsidized the mechanization of agriculture during the 'Green Revolution' in the 1960s and 1970s, which enabled many large farmers to displace sharecroppers (Alavi 1973). The 'Green Revolution could have had a much more positive impact on alleviating rural disparities, if underlying structural problems such as the uneven landholding patterns had been addressed first, and poorer farmers facilitated to adopt more innovative technologies (Alavi 1973; Gazdar 2009). However, the World Bank and the Government of Pakistan have instead opted to rely on corporate farming as the source of investment and adoption of agricultural technologies needed to boost growth. The potential threat of capital intensive corporate farming diminishing sharecropping opportunities for poor farmers, by leasing large tracts of land from surrounding landowners, has not been paid much heed.

c. How the Tarakee Farm manages its human resources

How and why corporate farming can have an evident adverse impact in terms of diminishing land access for poor sharecroppers has already been established. Now an attempt is made to assess the nature of employment opportunities offered by corporate farming to the rural poor, many of whom belong to households which now no longer have access to land due to the expanding operation of the large corporate farm in their vicinity.

Overall, there was a sizeable workforce employed by the Tarakee Farm, comprising of permanent employees, seasonal workers, and daily wage labourers. However, the terms and conditions under which this workforce was employed varied significantly. Based on interviews with the human resource management team, as well as different categories of the workforce employed at the farm, this section demonstrates the variance in terms of employment terms for different categories of employees at Tarakee Farm.

There were over 400 permanent employees working at the Tarakee Farm during the time that I was staying at the farm. These included high to mid-level managers, professionals including agronomists, irrigation specialists, livestock specialists, and financial and marketing experts. In order to attract these qualified professionals, the Tarakee Farm offered them lucrative salaries, and provided them spacious housing, a range of other recreational facilities, and free transportation for their children to attend well-known private schools in the adjoining district of Bahawalpur.[28]

Besides several low-ranking support staff working as permanent employees (such as guards, drivers, office assistants, kitchen helpers, etc.), there were only a few dozen farm workers employed on a permanent basis. These permanent on-farm employees were working on the maintenance of irrigation channels and other rotating duties. These low ranking on-farm employees were not well remunerated, even with their permanent employee status. The monthly pay for a permanent labourer ranged from Rs 3,700 to Rs 8,000 per month, after several years of service. I interviewed a permanent farmhand who had been working as a senior water-mate for the past 25 years,[29] and was drawing a monthly salary of Rs 7,500. His salary the year before, prior to a raise, had been Rs 6,000. As a permanent employee, however, he was provided with a small living quarter comprising two rooms. His wife and children also lived on the Tarakee Farm premises, but they had to purchase their own food. Another permanent farm employee at the Tarakee Farm, who had been working as a farm labourer for the past ten years, was also provided a living quarter, but his pay was just above Rs 4,000.

All permanent employees were hired using a standardized procedure. Permanent staff members were also being provided bonus payments, which translate into an extra month's pay for permanent labour, when the farm achieves a bumper harvest. During the preceding five year period, the farm had given four bonuses to its permanent employees. The benefits of such profits were, however, not passed on to contract or daily wage labourers, which actually formed the bulk of the workforce at the Tarakee Farm.

Depending on varying inter-seasonal harvesting requirements, the farm employs somewhere between 200 to 400 seasonal workers, who work on a contract basis for fruit picking at its mango orchards spread over an area of 800 acres. The mango orchard in-charge informed me that he works with a limited number of contractors, who in turn bring in workers from nearby districts in southern Punjab. One of these contractors was present at the time of my field-visit, and he informed me that he brought up to 300 seasonal workers to the Tarakee Farm on a yearly basis. This contractor said he brought in seasonal workers using three *jamadars*.[30] Besides working in the orchards, these *jamadars* were being paid an additional Rs 1,000 per month for supervising the work of their team of seasonal workers. The pay scale of this seasonal workforce was, however, modest. Fruit pickers working on Tarakee Farm were paid Rs 4,500, fruit graders got Rs 6,000, fruit packers got Rs 8,000, and fruit fixers (who were responsible for closing up the mango crates) were paid Rs 5,000. The Tarakee Farm was bearing the cost of the transportation of the workforce at the beginning and at the end of the mango-picking season, and it allowed the workers to sleep on its land, and provided them meals as well. However, provision of such basic facilities was a common practice amongst most large landowners when they hired seasonal contract workers. The Tarakee Farm did not offer any other incentives to these seasonal workers, such as the bonuses that senior management or other permanent employees receive.

The farm management had put in place a policy of paying seasonal workers individually, rather than through the intermediaries, so as to prevent potential conflict between them. All seasonal workers I spoke with appreciated this policy, as it minimized the chance of intermediaries taking a share of their pay. However, these seasonal workers mentioned that they had to leave their families behind in their own districts, and spend several months of the year working in other districts,[31] since they did not possess sufficient land for cultivation to ensure their households' survival. Although many of these seasonal workers had been coming to the Tarakee Farm for years, they did not have any written contracts, or any form of employment security other than maintaining good ties with their local *jamadars*. The Tarakee Farm management was evidently dependent on seasonal workers to meet its work requirements during the crucial harvest-time phase of production. It did not however develop direct contracts with the seasonal workers, and preferred to deal with intermediaries instead, who are held responsible for managing this workforce, without the Tarakee Farm requiring any formal interaction with them except paying their salaries on time.

Corporate farms are described as being particularly keen to seek out a female workforce to produce agro crops cheaply, by taking advantage of gender biases within the marketplace (Deere 2005). My research illustrates how the Tarakee Farm also relies heavily on a female workforce. In fact, my findings further reveal that a major factor compelling women to work at the corporate farm is due to the other associated impacts of corporate farming, such as the growing scarcity of land available for cultivation in the surrounding area. The fact that the Tarakee Farm itself does not offer poor farmers sharecropping opportunities further compels their women to work on a daily wage basis despite the meagre remuneration.

Up to 1,000 women and under-aged girls[32] were said to be working at the Tarakee Farm as daily wage workers, on average, which was over twice the number of people employed by the farm on a permanent basis. Given the size of the Tarakee farm, and the variety of crops being grown by it, daily wage work for women was available around the year. Female daily wage workers were found to be doing a variety of jobs, including cotton sowing, weeding, mango harvesting, orchard weeding, cotton picking, vegetable transplanting, and vegetable picking. Women and girls employed on daily wages worked eight hours a day and got paid Rs 85 per day for their labour. This amount totals up to Rs 2, 550 per month, given that they work all 30 days of the month. For cotton and vegetable picking the pay is marginally higher, since the payment for this task is made on the basis of how much of the crop they manage to pick.[33] If additional labour was required on a given day, daily wage workers were also paid for extra-time; if they put in another four hours (a total of 12 hours of work), for example, they could claim a day and a half of wages. All the women and girls employed on Tarakee Farm were found to be working on salaries not only below the official minimum wage in Pakistan, but also below the market rate for daily agricultural wages in surrounding communities.[34]

Farm supervisors for different crops had been appointed to oversee the work of daily wage earners. Most of these supervisors were also local villagers, and they were responsible for gathering the required number of daily wage labourers from surrounding villages within a 15-km vicinity. The Tarakee Farm used tractor-trailers to pick up dozens of women and girls for work from centrally designated collection points. A time-sheet system was maintained by the supervisors, according to which the daily waged women were paid every 15 days. There was no other paperwork acknowledging the affiliation of this female labour force with the Tarakee Farm, even though many of them have been coming to work here for years.

Cotton picking is traditionally one of the major agricultural tasks performed by women and children. Cotton pickers in this particular district were being paid Rs 49 per *maund* (40 kg) for cotton-picking in 2007 (Hisam 2007). Due to inflation, and the rising prices of cotton in general, these rates have gone up significantly. However, at the Tarakee Farm, women were being paid Rs 100 per *maund* (40 kg) to pick cotton, and they could usually pick no more than two *maunds* per day according to supervisors. However, this remuneration was much less than that being offered in surrounding areas. According to a female activist affiliated with a local community-based organization, women in the adjoining tehsil of Dunyapur had launched a successful campaign to get paid Rs 300 for picking one *maund* (40 kg) of cotton.

The Tarakee Farm supervisors did point out incentives for female daily wage workers to work for the corporate farm instead of landowners in the surrounding vicinity. One of them mentioned that women were been allowed to take away grass growing in the Tarakee Farm's mango orchards for free, after they are done with work, which can be used as fodder for livestock. A labour supervisor pointed out that some female labourers also received *zakat* payments,[35] depending on need and the duration of their employment at the Tarakee Farm. Another benefit included provision of free fortnightly lunches for all labour working on the farm, including contract and daily wage workers.

A senior labour supervisor at the Tarakee Farm informed me that until three years ago, they used to pay Rs 50 per day as daily wages, which was well above the daily wage rate of Rs 30 per day in surrounding areas. This rate was revised on a six-month basis, and had now come up to Rs 85 per day. However, local activists refuted the Tarakee Farm's claim to be paying better daily wage rates to its workers.[36] Other labour supervisors at the Tarakee Farm also agreed that they paid less than the market rate for daily waged agricultural workers. It seemed puzzling that the Tarakee Farm was paying these women less than the prevalent daily wage market rate, yet they still chose to work there. A senior human resources manger tried to explain why this happens. He highlighted incentives due to which women prefer to work at the Tarakee Farm. I was told that even if landlords in adjoining areas agreed to pay Rs 100 or more per day, this was still not a very attractive option for women daily wage earners, since they knew that these other landlords would employ them only for a few days, whereas they were assured work year round at the Tarakee Farm. The female daily wage workers I interviewed also confirmed that they could get better wages working elsewhere, ranging between Rs 150 to Rs 200 per day. Yet many of them admitted that that work on other farms was inconsistent and that other landlords promised them a certain amount but often did not pay them on time, or would try to negotiate and lower the rate once the work had been completed. While working at the Tarakee Farm provided them greater income security, the actual salary made by these poor agricultural workers for putting in an entire days work was much below the official minimum wage in the country.[37]

I was taken to visit a government basic health unit located on the Tarakee Farm, the diagnostic facilities of which had been upgraded (ECG, ultrasound) through funding provided by the corporate farm. The provision of facilities such as primary health care to its workers may be serving as an indirect incentive to attract a substantial workforce despite inadequate remuneration.

The senior management was planning to formally register the Tarakee Farm as a corporate entity, which would imply the need to comply with certain labour welfare obligations, such as providing the minimum wage and insurance facilities to its permanent employees. However, daily wage earners or contract workers would remain informally employed, and thus not be able to avail these facilities.

A few of the daily wage women working on the Tarakee Farm mentioned that their families owned small pieces of land ranging from a few *marlas* (there are 20 *marlas* in 1 acre) to about 2 acres, but they still had to work to help meet rising household expenditures. Most of them, however, did not own any land. Half a dozen of them also mentioned that their husbands work as labourers in the off-farm sector, since there were now much fewer sharecropping opportunities available to them. However, the labour supervisor overseeing the work of these women later told me that most of the female respondents I interviewed had husbands who were drug addicts or gamblers, and blamed these traits for their inability to work rather than the lack of sharecropping or lease farming opportunities, which have been compounded by the large-scale operation of Tarakee Farm itself. However, none of the dozen or so women that I had discussions with, mentioned that their husbands had substance abuse or gambling problems.

In one of the agricultural plots on the Tarakee Farm, I also noticed around four girls under the age of 14 working alongside dozens of older girls and women, which is technically considered child labour.[38] When I asked the senior Tarakee Farm managers if they had a differentiated pay scale for younger girls, I was informed that they did not discriminate on the basis of age, even though younger girls cannot work as hard as mature women. Since younger girls received the same rate as older women, this incentive seemed to encourage, rather than deter, child labour. Although I was at the Farm during school summer vacations, no explicit rule was identified by the management which would discourage young girls from working in the field instead of going to school even during the school year.

While working in cotton fields is not listed as a hazardous industry, Pakistan has come under repeated criticism by the international community, and international and national child rights groups, for tolerating child labour in cotton farming (Murray and Hurst 2009; Society for the Protection of the Rights of the Child (SPARC) 2007). The prospect of promoting corporate farming does not seem to be discouraging this trend.

On the whole, the Tarakee Farm had provided very limited prospects for the permanent employment of poor farmers from surrounding areas. Instead, it was primarily relying on contract and seasonal labourers, and daily waged female labourers. While tractors and harvesters are being used on the corporate farm, I was also informed by the senior management that they were going to begin investing more heavily in machinery to further lessen their dependence on contract and daily waged labourers. The detrimental impact of shifting to more capital-intensive farming on employment opportunities for local farmers was not an issue of concern for the corporate farm management.

Despite the confidence shown by donor agencies in the ability of corporate farms to increase employment opportunities in rural areas, the above findings illustrate that the opportunities for work emerging on a leading corporate farm within Pakistan are limited. Leaving aside the well-incentivized top management, the employment available at the corporate farm for poor farming communities are confined to low waged positions for female workers, or for seasonal workers. Moreover, the terms offered to these low waged workers are not commensurate to wages available in surrounding areas. The corporate farm seemed tolerant towards exploitative practices such as child labour, and there was no evidence of it ensuring the safety of its workers employed to pick pesticide-sprayed cotton, despite its documented ill effects on health. The ability for such corporate ventures to alleviate poverty and contribute to the overall development goals of empowering the rural populace, therefore, seems very limited.

Further observations concerning Tarakee Farm

The Tarakee Farm had visibly undertaken many innovations, grown steadily, and been able to draw the attention of development agencies, including international donors, and seems poised to work with multinational and other large corporations to further boost its profitability.

The Tarakee Farm had further managed to win the favour of bilateral development agencies like USAID, which helps it export its mangoes to mainstream European markets.[39] Project documents for a USAID initiative claimed the objectives of creating greater economic opportunities in the agricultural sector by sending premium quality mangoes to buyers in the EU instead of cheaply selling their produce locally (FIRMS 2010). However, a major beneficiary of this intervention was a corporate farm rather than poor farmers. Moreover, while exporting mangoes to the EU meant a boost in corporate profits for the farm, none of the daily waged female labourers, or the seasonal workers brought in to pick and package the fruit for export, were being provided any extra remuneration for their work.

The Tarakee Farm remained keen to link itself to other multinationals and wholesale businesses, in addition to creating further inroads into the export market. The Tarakee Farm was still selling 50 per cent of the milk produced by 1,000 Australian cows on the farm, and 70 per cent of its vegetables, to wholesalers. However, it aimed to increase linkages with multinational and wholesale business houses. It had begun supplying milk to multinationals like Nestle, and was trying to explore selling its produce to Makro-Habib,[40] a leading wholesaler of food and non-food products with an established network of outlets in the metropolitan cities of Lahore and Karachi. The Tarakee Farm's strategic decision to work with other large corporations may work well in terms of boosting corporate profits, but the impact of cutting out local intermediaries, or potentially driving up the prices of food items in local markets due to its plans to send more of the locally produced crops to other cities and abroad, cannot be discounted.

It was interesting to note that the Tarakee Farm was giving a significant amount of money to the government in taxes. These taxes were being calculated on a self-assessment basis, given that the Agricultural Income Tax department is still primarily collecting revenue on the basis of landownership rather than being able to assess the agricultural productivity of land. While this is indeed a noteworthy effort, given the rampant tax evasion in the agricultural sector, the fact that the Tarakee farm gave money to the federal government did not directly translate into benefits for those farmers who are being adversely impacted by its operations. While the corporate farm provided an opportunity for larger landowners to enter into long-term lease arrangements, poor farmers were mostly being marginalized by the presence of this large corporate farm operating over a vast amount of the agricultural land.

An economic historian and agri-business expert recognized that the Tarakee Farm owner has the strategic advantage of ensuring increased agricultural capability, and this business model is also poised to help boost exports. However, this business model is clearly not transferable to smaller farmers. The Tarakee Farm corporate farming model instead seems to be reverting to the same elite-led farming model adopted by the state and donors at the time of the 'Green Revolution', despite ample evidence that such modes of agricultural productivity strain the equity equation at the cost of income optimization. Several adjoining large landowners had been trying to replicate the Tarakee Farm model. However, no one else was said to have been successful in turning their operations into such a professionally run and profitable venture. Besides the hard work of the senior management team, a large part of the success of the Tarakee Farm business model was attributed to its owner, who had been able to leverage state and donor resources (such as stabling a basic health unit on its premises, or securing support for exporting its produce to foreign markets) to supplement its business model.

While some of the literature on class and society in Pakistan identifies the military, bureaucracy, capitalists, and landlords (large landowners) as distinct categories comprising the ruling class of the country (Herring 1983; Ahmed 1996), there is also some recognition of the crosscutting nature of such categorizations. For instance, large landowners have been known to use their control over the rural populace to assume political power, or else, they have diversified in terms of their capital-generating capabilities and became industrialists and entrepreneurs as well (Gardezi and Rashid 1983). It is this ruling class that has been able to adapt itself easily to new forms of production, such as corporate farming, which fit in well within the donor-supported rural development agenda. Yet the corporate farm does no more than offer a modest range of under-remunerated employment opportunities to poor farmers.

Conversely, findings emerging from the Tarakee Farm illustrate how corporate farming is actually exacerbating land scarcity, and placing an increasing burden of household survival on women. Such findings are consistent with critiques of corporate farming emerging from other developing countries in Africa, Latin America, and Asia (Lorenzo *et al.* 2009; Pereira 2005). Still, these adverse impacts have not yet dampened the enthusiasm expressed by governments of countries like Pakistan, and international agencies like the World Bank, to exercise more caution in encouraging corporate farming as the means to bolster agricultural productivity and to alleviate rural poverty.

Conclusions

The need to achieve growth to alleviate poverty remains vital. However, the donor community justifies, and is primarily supporting, market-led and growth-based solutions to agricultural development and rural poverty alleviation. It does so without placing due emphasis on the need for equitable distribution of productive assets, which, in the context of rural Pakistan, requires addressing structural inequalities such as the skewed ownership of land as a basic prerequisite. Failure to address such issues results in the failure of donor interventions to address the challenges facing poor farmers.

Within the context of the increasing liberalization of the agricultural sector in Pakistan, this chapter has drawn on field research to illustrate how the underlying political economy of landownership and power relations in Pakistan are, in fact, causing a range of unexpected market distortions, and creating service gaps which are ushering in profit-driven multinationals to gain more control over the provision of agricultural inputs and extension services to the detriment of poor farmers.

An analysis of salient donor-endorsed policy mechanisms, like the PRSP, indicates no more than a rhetorical admission of the land inequality problem, and an emphasis instead on the provision of micro-finance to the poor and using agri-business to help boost growth. While the PRSP-endorsed programme of computerizing land records claims to have explicit poverty alleviation goals, such as lessening the burden of litigation and preventing the abuse of poor farmers by corrupt land revenue officials, the thrust of the existing LRMIS programme in Punjab is focused on computerizing existing land records, rather than correcting them, or using the land records to undertake any redistributive scheme. While such computerization efforts may facilitate expedient land transactions, and help large landowners to lease lands to capital intensive agri-businesses, it is difficult to see how it would translate into better outcomes for poor farmers.

Another donor-endorsed approach of using NGOs to help poor farmers gain access to credit or land markets was also considered and found to be inadequate in terms of offering poor farmers viable opportunities. Findings emerging from research in Sindh, where mainstream micro-finance entities are working, indicate how some poor farmers are becoming more indebted due to the provision of credit, rather than being empowered. Another donor proposal designed to increase land access among the landless poor by enabling them to enter the land tenancy and land rental markets, is also problematic, since it mimics a cost-sharing and risk-sharing landlord–tenant relationship approach. Such an approach would also invariably result in exploitation of poor farmers by passing on the burden of intermediary costs onto them, without offering poor farmers any prospect of securing ownership of their own piece of land.

Donor emphasis on corporate farming as a strategy to achieve agricultural growth and poverty reduction were also explored with reference to the workings of a leading corporate farm in Pakistan. It is interesting to note that the owner of the Tarakee farm combines in himself political influence and entrepreneurial skills, which make him the ideal candidate from amidst the landed rural elite to instigate corporate farming in the country. The Tarakee Farm is praised by donor agency representatives and senior government officials. On ground research, however, revealed that while the Tarakee Farm is indeed an innovative and profitable venture, yet its remuneration-based incentives remain limited to the senior management. Contract and daily wage workers, in particular, are getting little benefit from the success of this business venture.

Instead, the corporate farm has indirect detrimental effects on poor farmers, such as exacerbating land scarcity for smaller farmers and driving up lease rents for those who cannot afford to pay advance payments for lease arrangements over multiple years. The corporate farm is also not interested in sharecropping, and provides little opportunity for poorer farmers other than working as daily wage labourers or contract workers. So, while similar agribusinesses may spur agricultural growth and corporate profits, they offer meagre prospects for poor farmers, and in fact may very well exacerbate their existing marginalization, despite optimistic claims made in donor-endorsed poverty reduction strategies for Pakistan.

Given the findings of this chapter, it is difficult to justify the utility of market-based solutions to achieve equitable agricultural growth without first addressing the underlying causes of deprivation resulting from unequal land-ownership patterns. Yet this is in effect what donor efforts continue to do, in order to perpetuate the liberalization agenda, even though the benefits of many of their above-described policies and programmatic interventions continue to bypass the poor and instead help preserve and further consolidate the position of the landed rural elite.

While donor influence on the Pakistani state has not helped alter the underlying structural impediments confronting poor farmers, there is evidence of more direct exertions of human agency by poor landless farmers to alter their existing circumstances, which is taken up in the following chapter.

Notes

1 The Indus basin comprises six rivers and the treaty gives India exclusive use of all of the three eastern rivers and their tributaries before the point where the rivers enter Pakistan. Similarly, Pakistan has exclusive use of the western rivers. While this treaty has remained in effect despite three major wars between India and Pakistan, it has come under pressure due to Indian dam construction projects. Other critics of the treaty point to the need for an integrated approach to the management of the Indus water basin, rather than the mere bifurcation of its waters.

2 A USD 90 million USAID Agribusiness Support Programme is one example of an ongoing five-year-long nationwide project specifically targeting export-oriented agricultural production.

3 The Post-Washington Consensus was a reaction to the negative impact of structural adjustment policies on inequality and poverty around the developing world, and the term implies an apparent shift to enhancing participation and country ownership of development strategies. However, the Post-Washington Consensus has also been criticized for placing a continued emphasis on neo-liberal policy prescriptions.

4 The prevailing land revenue system is manually maintained and managed by low-ranking officials called *patwaris*, whose role has already been discussed in earlier chapters.

5 Agricultural technology can include the introduction of new crop varieties and chemical products such as pesticides and fertilizers, improved management practices relating to crop or water use, and further mechanisation in agriculture or irrigation (such as the use of drip irrigation).

6 The NHDR categorized the rural non-poor as having landownership exceeding 6 acres. For the extremely poor and the poor, land ownership was under 1 acre and 2 acres, respectively.

7 Union Councils form the lowest tier of local governments within districts. There are 120 districts in Pakistan.

8 While originally Hindus, this community has since converted to Islam.

9 Large landlords often provide credit to their tenants or sharecroppers for the purchase of agricultural inputs, and for consumption purposes, in many other parts of the country as well.

10 Sharecroppers working as seasonal labourers were paid in kind for the wheat crop, 1.5 *maunds* (or 60 kg per family) per acre of wheat harvested, which took an entire family four or five full days of work. For cotton picking, local landlords offered Rs 200 per *maund* (40 kg), which required several hours of hard labour by an entire family.

11 While BT cotton is already available in the informal sector, multinational corporations such as Monsanto are under negotiation with the Pakistan government to sell branded BT cotton provided the government agrees to begin penalizing farmers found to be using their seeds without purchasing them directly from the company.

12 Protecting international property rights in agriculture in particular implicitly acknowledges the need to safeguard interests of agri-business concerns keen to promote genetically modified crops into the country.

13 This Urdu phrase literally means 'a new dawn'.

14 Mutations imply officially recording (in revenue department records) the transfer of title of the given property from one person to other(s).

15 *Gardawars* rank as being the immediate supervisors of *patwaris* in the land revenue system.

16 Lowest level of local governance comprise a cluster of villages and towns. Gujrat district has 65 union councils.

17 Manor.

18 *Numberdars* are villagers (often prominent landowners) selected by the revenue department to assist with revenue collection and maintenance of records.

19 *Tehsildars* hold a higher position than the *gardawars*, since they supervise land record management at the tehsil level, the second tier of governance within a district.

20 DfID is the United Kingdom's bilateral aid agency.

21 The term *taccavi* refers literally to agricultural loans or advances. The tradition of providing such short-term loans to farmers can be traced back to the Dehli Sultanate (1206 to 1526) in times of famine, and to subsequent rulers of the Indian Subcontinent. After independence in 1947, the Pakistan government also provided these short-term loans to farmers for the purchase of livestock, machinery, or other agricultural inputs such as pesticides or fertilizers. Provincial governments subsequently consolidated various laws relating to the advancement and recovery of *taccavi* loans for agricultural purposes. In the case of the Punjab, this effort led to the formulation of the Punjab Agriculturalists Loan Act of 1958 (Government of Punjab, 1958).

22 While micro-credit primarily focuses on the provision of loans, micro-finance includes a broader range of financial services, including the provision of health coverage or crop insurance schemes. Both micro-credit and micro-finance programmes aim to provide small amounts of money and charge high interest rates due to the high transaction costs associated with providing money at the doorsteps of the poor.

23 The National Human Development Report for Pakistan (2003) ranked this particular district where the corporate farm is located among the last five out of 29 districts in the Punjab on the basis of human development indicators including literacy, health, income levels, etc.

24 The actual number of seasonal and daily wage labourers varies depending on the agricultural cycle and yields of particular crops.

25 The Tarakee Farm owner's brother-in-law belongs to a renowned political family, which has held prominent government positions within the colonial era, and thereafter in Pakistan.

26 Cost-sharing strategies imply asking local communities to share in the burden of development works including provision of education, health, or sanitation facilities, and are endorsed by donors such as the World Bank as the means to ensure community participation, while ensuring that public sector expenditure does not become inflated. Critics of this approach however consider such development strategies unfair for passing on the public responsibility of basic service provision onto already marginalized communities.

27 Ayub Khan was the first military ruler of Pakistan.

28 The distance between the two districts is only 17 kilometres, which makes it easy for the children of the higher management to commute to private schools on a daily basis.

29 The work of water-mates implied undertaking manual labour to regulate and channel the water flow in the on-farm watercourses. This senior water-mate also had several assistants working under him, their number varying from three to over a dozen depending on the irrigational needs of different crops.

30 *Jamadars* usually collected workers from their own community and took them to work seasonally in areas where seasonal labour is required. They often worked alongside other contract workers but got a commission for identifying work opportunities, and also dealing with the employers on behalf of the seasonal workers. *Jamadars* can work independently or under the supervision of a larger contractor, as was the case at the Tarakee Farm.

31 In addition to coming to southern Punjab for the mango-picking season in the summers, many of these seasonal workers also went to Sargodha, a district in central Punjab, for orange-picking in the winters.

32 The Employment of Children Act 1991 categorizes the minimum working age to be 14 years but it does so for a range of 'hazardous' sectors which do not currently include the work of cotton picking, although several NGOs have labelled such work as being hazardous as well.

33 In the case of vegetables, women on average were said to pick between 40 to 50 kilograms per day, and they were paid between Rs 2 to Rs 3 per kg depending on the vegetables being picked, but this work was only available seasonally.

34 The minimum wage in Pakistan during the time of this fieldwork in 2011 was Rs 7,000 per month, albeit a vast proportion of the country's workforce is paid below the minimum wage, especially in the non-formal sector, which includes agriculture.

35 Based on the Islamic notion of alms giving to help those without kin, although rates vary, the generally accepted *zakat* rate comes to 2.5 per cent of annual income. The government of Pakistan passed a law to collect *zakat* (the *Zakat* and *Usher* Act) in 1980.[m1]

36 A local female activist also claimed that she had helped wage labourers organize in the adjoining union council to demand and receive daily wages up to Rs 150 per day.

37 The minimum wage in the country was Rs 7,000 per month at the time that this research was conducted, while the women working on Tarakee Farm were getting paid half that amount. While it is relevant to note that the official minimum wage is applied mostly to the formal sector, given the fact that the Tarakee Farm claims itself to be the most professionally run farming venture in the country, there is no reason why it could also not adopt minimum wage standards prescribed by the government.

38 The ILO's Minimum Age Convention (C 138), to which Pakistan is a signatory, makes a distinction between working children and child labour. Child work can be in non-hazardous industries after children reach the age of 14 years, provided they have working hours and are able to continue going to school, whereas children working under the age of 14 is illegal (ILO 1973).

39 The said project involves a multimillion dollar commitment over a five-year period aiming to develop an internationally competitive business sector in Pakistan, which focuses on increasing exports and producing higher value added products and services as the means to enhancing economic growth and reducing poverty.

40 Makro-Habib is a relatively new business venture by the established business house of Mohammed Ali Habib Business group, whose investments include prominent banks and insurance businesses in the country.

6 Resisting and challenging adversities confronting poor farmers

The inability of state institutions to transcend elite-focused agricultural development policies, and international donor agencies like the World Bank's emphasis on propagating the market mechanism, continues to compound structural problems confronting poor farmers. This situation has provoked instances of resistance by poor Pakistani farmers themselves to alter the status quo. This chapter will first point to the long history of resistance movements involving farmers in the Indian subcontinent, and then identify and discuss a recent prominent peasant movement in the Punjab, the Anjuman Mazareen Punjab (AMP).[1] After highlighting the unique characteristics of this local resistance movement, launched by tenants threatened by eviction from military-owned farms, some of the challenges confronting the AMP are identified. While acknowledging the agency of landless farmers to stand up for their rights, the problem of increasing fragmentation and myopia in the AMP, and the contested role of NGOs/INGOs in supporting the AMP will be discussed, which are major stumbling blocks undermining the potential ability of the AMP to challenge military control over agricultural lands in the country.

Resistance by farmers

Struggles over access to, and control over, land, as well as resources located within those lands (including water, forest areas, or other minerals), remain the source of violent conflict in different parts of the world (Borras and Franco 2010). Eqbal Ahmad (1980) points to how peasants in the twentieth century have been the primary instruments of revolution in a diverse range of countries such as China, Cuba, Algeria, and Vietnam.

While the recent AMP movement in Pakistan has generated a lot of interest among activists, campaigners, and academics, historically peasant movements have taken place repeatedly in rural areas that are part of Pakistan today. Consider, for instance, the decade-long guerrilla resistance in the Punjab sparked by Dulla Bhatti, and his family's refusal to pay land revenue during the zenith of the Mughal Empire (Salim 2008).[2] During the latter part of the seventeenth century, the *Sufi Sheikh* Shah Inayat also struggled for establishing a more egalitarian rural society in Sindh.[3] The increasing extractive pressure applied by the Mughals on the peasantry during its declining years evoked strong resistance from the peasant-soldier communities of the Marathas, Sikhs, and Jats, which served to hasten the downfall of the Mughal Empire (Habib 1983; Salim 2008).

The early years of British rule in India were also marked by widespread peasant rebellions. Long before the rebellion of 1857 – which is often described as the first sign of a struggle for Indian independence – it was the peasants of Bengal and Bihar, who revolted against the East India Company and its growing land revenue demands, amidst famine-like conditions in the 1770s. Another uprising of peasants across south India in 1800 succeeded in forming a Peninsula Confederacy, in defiance of British rule, which required colonial forces to be requisitioned from Ceylon, Malaya, and even from the UK itself, in order to crush this rebellion (Habib 1983, 1999; Rajayyan 1971; Salim 2008).

During the 1857 uprising as well, which was instigated by a mutiny of *sepoys* (soldiers), a plethora of peasant bands had formed all over north India to participate in the uprising, except in places like the Punjab where land grants and military recruitments had helped create compliance among the peasantry (Ali 1988). Soon after the 1857 rebellion was crushed, the peasants of Bengal, forced to cultivate indigo for British planters to cater to the growing textile industry in the UK, revolted in 1859. The Pabna peasant uprisings of 1872–3, in what is now the Sirajganj district of Bangladesh, included both Hindu and Muslim leaders as well as peasants. The creation of the Indian National Congress in 1885 and the Muslim League in 1906, which spearheaded the independence movements for India and Pakistan, were also catalyzed by such peasant movements (Stokes 1978; Salim 2008).

Eqbal Ahmad's (2007) differentiation of peasant movements, as being either rebellious or revolutionary, provides relevant insight into categorizing peasant movements in countries with a colonial legacy. In pre-capitalist societies, peasant groups are described as having been composed of a multitude of small, mutually exclusive groupings, whose myopic ideologies deterred their integration into a larger social movement. This lack of integration had important consequences in terms of limiting peasant cooperation across the boundaries of clan and community. Hence, local problems rarely emerged to become society-wide issues, and while peasants rebelled against excesses, they did not question the legitimacy of the larger system within which they were situated. Karl Marx noted this phenomenon as well, while describing the lack of revolutionary potential in the French peasantry (Marx 1963: 163). To Ahmad, however, it was colonialism which became a catalyst for a transition in the role of the peasant movements in the Indian subcontinent, by imposing a state system which differed radically from the pre-capitalist monarchical, feudal, and tribal systems of power. The consequent demographic and socio-cultural disruption was further compounded by the commercialization of rural communities, in turn exacerbating a crisis of authority. Unlike pre-capitalist rebels, Ahmad asserts that peasants in formerly colonized countries thus began to seek overthrow, rather than renovate, the prevailing social order. The process of colonial subjugation is described as causing peasants to begin viewing themselves as part of a larger society, and to start identifying with other marginalized groups against a wider system which was seen to be directly responsible for their personal deprivation.

The peasantry in the Indian subcontinent did become increasingly politicized as well, and created a unified platform to voice their concerns, by establishing the All India Kissan Committee in the mid 1930s. This entity formulated a charter aiming to address specific poor farmers' concerns,[4] and to also contribute to the nationalist movements aiming to secure independence from the British. Yet while the Kissan Committee formed provincial chapters, large landowners also became increasingly active in politics, especially in provinces like the Punjab (Salim 2008). The Muslim League sought support from both the Kissan Committee and large landowners to help strengthen its demand for an independent homeland for the Muslims of the Indian subcontinent. After the partition of the subcontinent had taken place, however, the ruling Muslim League decided to increasingly adopt growth-led agricultural growth strategies to circumvent the need for redistributing assets, especially land, owned by politically affluent landowners to the landless poor, as discussed in the preceding chapters.

Ahmad (1980) had hoped that as large numbers of people previously outside the political realm forced their way into national politics, the common man would become increasingly relevant to politics. This, unfortunately, did not happen in the case of Pakistan. Instead, large landowners, a narrow capitalist class, alongside the bureaucracy and military, managed to dominate political developments in newly created Pakistan.

In 1968, an attempt was made to take a consolidated stance against the landlord and capitalist elites by forming the Mazdoor Kissan Party (MKP).[5] However, two years later, when democratic elections were held after a long period of military rule, the MKP was not sufficiently organized to participate in the general elections. The newly elected PPP government tried to implement land reforms in 1973, but the fact that its senior leadership itself belonged to prominent feudal families meant that these reforms produced lacklustre results (as discussed in detail in Chapters 2 and 4). Yet when elections were held again in 1977, the MKP remained unable to contest. In the subsequent three decades of military rule and democratic governments, the MKP has not managed to become an effective political party either. It has fragmented into different factions, which have merged with, and then disassociated themselves from, other leftist groups like the Communist Party and the Pakistan Workers' Party.

During an interview, the Secretary General of the MKP admitted that their party has not been able to make its presence felt on the political landscape of the country, but he pointed out how the MKP continues working at the grassroots level with peasants, especially in the North West Frontier Province (NWFP), now known as the Khyber Pakhtunkhwa Province (KPK).[6] Most recently, the MKP has been supporting tenants against landlords trying to evict them in Hashtnagar in the Charsadda district of NWFP/KPK.

The Labour Party of Pakistan, formed from the reorganization of earlier leftist groups in 1997, has also tried to support resistance movements. It has done this by helping form the AMP in Punjab and by supporting the Pakistan Fisherfolk Forum (PFF)[7] along the coastal areas of the provinces of Sindh and Baluchistan. However, the role played by the politically insignificant leftist parties in supporting social movements in not without contention, as is discussed in the following sections in more detail.

The Anjuman Mazareen Punjab and its stand against military farms

In comparison to other peasant revolts, the AMP's resistance is considered particularly significant since it has managed to resist the most powerful institution of the state in Pakistan, the military. Before assessing its significance, the broader context within which the AMP and its resistance to military farms emerged needs to be mentioned.

While the AMP movement eventually spread across several districts of Punjab,[8] it was in Okara district, where the AMP originated, and where its resistance proved to be most rigorous. Geographically, the Okara district is located in the Lower Bari Doab,[9] which was part of the area where colonial irrigation and land grant schemes were initiated during the late nineteenth century (Akhtar 2006; Ali 1988). Since vast tracts of agricultural land became available due to the irrigation schemes, the British government allocated land grants to create a compliant class of peasant proprietors, and win over the support of local influentials. Land grant schemes were also used to further military aims, such as breeding for cavalry regiments, and to serve as inducements and rewards for military recruitment and service, respectively. The distribution of land in Okara served this latter objective of supporting military concerns.

Between 1914 and 1924, the colonial administration provided several land grants to the military for breeding horses, to establish farms for providing dairy products to soldiers, as well as to grow oats and hay for cavalry horses. The Military Farms Department received an allotment of around 20,000 acres to establish the 'Oat Hay Farm', which subsequently became the largest state-owned farm in the entire Indian subcontinent (Ali 1988: 34). There are two military farms in this area now; and the Renala and the Okara Military Farms. The latter is the larger one, which comprises 18 villages. The research presented in this chapter was conducted within the Okara Military Farms, at the Tiryana (Military) Estate (not real name), which itself consists of seven villages and three smaller hamlets.

The land for the Tiryana Military Farms in Okara was handed over to the British Indian Army for cultivation purposes, initially for a period of 20 years. This lease was subsequently extended, so the land remained under the Defence Ministry's use until the time of the Indian subcontinent's partition in 1947, when its possession was transferred to the Pakistan Army. The provincial government stopped receiving any further rent on this land after Pakistan was formed (Akhtar 2006), but the provincial authorities remained reluctant to confront the powerful army to demand rent, let alone try to evict the military farms' administration from land legally owned by the provincial government.

The colonial land settlement policies had preferred granting land to 'agricultural castes', and they, in turn, brought agricultural labourers (*kammis/ haris*) with them to help cultivate newly irrigated lands. While the agricultural labourers did not have landownership rights, the agricultural castes were usually issued inalienable property rights if they spent a specified amount of time working these lands, usually under 15 years after the initial settlement. On lands meant to be retained by the state (including the military farms), agricultural castes were encouraged to help with the cultivation. While land rights were also promised to agricultural castes cultivating state-owned agricultural lands, these rights often did not materialize. Thus, even agricultural castes working on state lands, such as the above-mentioned military farms, were cultivating these state lands as sharecropping tenants, rather than as land grant beneficiaries. These agricultural tenants were categorized as 'tenants-at-will' within revenue records, rather than as 'occupancy tenants', and the latter had more rights under tenancy legislation, even under British colonial rule. The tenants-at-will on state farms continued to be denied property rights even after independence was gained (Akhtar 2006; Ali 1988). Farmers on the Okara Military Farm had been trying to secure ownership rights even before independence took place and they have since filed court cases as well, but remained unable to secure a favourable outcome (Mumtaz and Mumtaz 2012).

The land where the Okara military farms are located has thus remained under military control, being used for growing animal fodder, and even cash crops, to generate revenues for the military establishment. Over time, however, due to fluctuating agricultural prices and the alleged corruption of the military farm supervisors, the profitability of these farms began to decline (Akhtar 2006). In an effort to increase profits, the farm management announced in 2000 that the sharecropping arrangement would be replaced by a new contract system, whereby tenants would give fixed cash rents rather than a share of their yield to the military farm. Military farm officials had justified this decision based on the sharp fall in annual farm income, from Rs 40.79 million in 1995–6 to Rs 15.87 million in 1999–2000 (ibid.: 489). It is also significant to note here the military farms' attempt to abandon more traditional forms of farming and adopt a contract farming model is indicative of the increasing commercialization of land tenure patterns, which donor agencies continue to promote, by encouraging use of market mechanisms within different aspects of agricultural production, including the land tenure system.

According to research describing the emergence of the conflict on the Okara Farms (Akhtar 2006; Bano 2012; Mumtaz and Mumtaz 2012), the imposition of a contract system undermined the land tenure security of both well-off and less well-off tenants-at-will who had been cultivating these state (military) agricultural lands since generations. Making sharecropping tenants pay an annual rent in cash instead, with the new contracts being subjected to annual renewal on the basis of economic performance, would effectively have enabled the military farm management to begin evicting tenants who grudged paying the demanded amount. When sharecropping tenants showed hesitancy in accepting these proposed changes, the military farm management began to displace non-compliant tenants, and encouraged non-village residents to sign lease contracts to begin cultivating the evicted tenants' land instead. These developments provided an impetus to the formation of the AMP resistance movement (Akhtar 2006; Arif 2008).

Being unable to reach a compromise agreement with the military farm management, sharecropping tenants began to express increasing levels of agitation. Initial protests took place at the end of 2000, when hundreds of tenants on the military farms went to protest at the Deputy Commissioner's Office in Okara (Arif 2008). The failure of the highest civilian government official in the district to effectively intervene in a dispute between landless farmers and the powerful army caused an escalation in the tensions.

The AMP's resistance movement initially aimed to reject the contract system of leasing, and to prevent the military farm management attempts to bring in contract farmers to replace local tenants who had been local residents for multiple generations. Yet, once the tenants became more organized and realized that the army itself was not the legal owner of the agricultural land in question, the AMP objectives evolved into a struggle for ownership rights. However, even though the military farms were themselves not paying rent on lands owned legally by the provincial government, the state machinery, including the police, supported the military farm management in helping suppress tenant protests, which became increasingly belligerent. The AMP activists I met pointed out that, besides repression through the state machinery, including the police and even Rangers,[10] the army management also began to bring in local gangsters to intimidate village residents. Irrigational and electricity supply to some of the villages on the military farms was also suspended. The ability of the army to compel other government institutions to exert pressure on the AMP resistance is an indication of the dominance of the institution within the prevailing political economy of the country. Moreover, when the villagers resisted this intimidation, also resorting to violence, they were in turn subjected to further repression. Some of the AMP activists accused the military farm management of lodging fabricated cases of terrorism against the AMP activists who also used force to resist attempts to evict tenants from the military farms.

To counter these varied coercive tactics being employed against them, the AMP began to encourage resistance from women residing in the villages located on the military farms. Initially women would try to physically shield their men from violent crackdowns against the protestors, but soon thereafter, women also began taking more active part in AMP protests and rallies within their own districts, and beyond. Besides allowing women to participate in the resistance movement, the AMP also used other pressure tactics such as hunger strikes, appeals to the media, and even to the international community to support their cause (Mumtaz and Mumtaz 2012; Zaidi 2012b). This wider strategy of resistance was adopted due to the involvement of several prominent NGOs, which had, by this time, also rushed in to lend support to the AMP.

Despite the consistent repression, within a few months, the AMP protests spread into nine other districts of Punjab, where landless farmers were facing similar eviction pressures as the landless tenants on the Okara military farms. In the face of multipronged resistance by the AMP, as well as national and international condemnation of the repressive tactics used by the state, the paramilitary troops (Rangers) were withdrawn in August 2003, and local tenants were able to begin re-cultivating the military farmland. Lauding this apparent victory, Akhtar (2006) noted that tenants had not surrendered harvest shares to the authorities, which was cited as an obvious indicator of the AMP's success. This enthusiasm was reiterated three years later, based on observations that villages at the epicentre of AMP activism were continuing their struggle by not only still refusing to give *battai* (a share of their produce), but by having instituted court cases against military farm management, and its undue use of force against them. Those who had been forced by the military farm management to sign lease contracts were also said to have also rejected their contracts and stopped paying lease rents to the military (Mumtaz and Mumtaz 2012). My own field research, however, revealed a different situation. I found evidence of non-resident army allocated lease contractors in villages on Tiryana Estate, as well as former tenants paying lease amounts to cultivate military farmland. The ability of the AMP to have altered the tenancy arrangements on the military farms, therefore, merits more attention.

The AMP's significance has also been emphasized by reference to its uniqueness when viewed from the lens of conventional Marxist theories of peasant revolt. Akhtar (2006), a political activist and researcher who worked closely with this movement, argues that the social structure on the military farms in Okara evolved very distinctly from other canal colonies, where different sizes of land grants were provided. In the case of the military farms in Okara, there was no stark class distinction between tenants since most of them had been allotted the same amount of land (typically 25 acres) for share-cropping purposes. While their landholding sizes did not remain constant, due to the varying number of heirs, the majority of tenants remained locked in sharecropping arrangements where they were receiving no more than one-third or even one-quarter share of the produce. Akhtar also argues that the authority relationships on the Okara military farms were much more pronounced between the military farm administration on the one hand, and all classes of landless farmers (i.e. tenants sharecropping on varied sized plots of land) as well as wage labourers, on the other. This diametric relationship enabled the formation of a united front of all landless farmers against a common adversary; the military farm administration.

Even caste and religious differences were said to have become insignificant in terms of developing a sense of political consciousness within the AMP.[11] Akhtar, therefore, builds a case for the AMP being a distinct category of peasant revolts, which has not addressed by the conceptual framework utilised in a majority of Marxist analyses.[12] He hails the AMP for having brought different classes, castes, and religious groups together onto a common platform. However, emerging tensions within the AMP movement, and the subsequent fragmentation it has experienced, undermines such an encouraging assessment.

The following sub-sections highlight some of the problems that have come to afflict the AMP at the Tiryana Estate, which have undermined its overall success. Three major problems are discussed, which indicate that the AMP lacks the ability to pose a serious challenge to the prevailing political economy which marginalizes poor farmers. The first of these problems concerns the inability of the AMP to neither secure legal landownership rights for tenants working on military-controlled lands, nor to prevent contract farming in AMP villages, despite its temporary success in halting evictions of tenants. The second problem relates to the increasing myopia of the AMP, which is undermining the much lauded ability of the movement to have initially transcended social boundaries within rural societies to take a unified and multi-pronged stance against the military. This problem is discussed with particular reference to gender-related biases, which now plague both the leadership cadres and the movement in general. Besides marginalizing women, the male domination of the AMP has also led to the emergence of an increasingly narrow-minded agenda, which is weakening the capacity of the movement to sustain credible resistance against the military. The final set of issues to which attention is being drawn concerns the problematic nature of the support provided by civil society organizations to the AMP, which has also exacerbated fragmentation within the movement, and also remained unable to link the AMP with broader social movements.

a. Lacklustre achievements of the AMP

The AMP has been credited with taking a bold and united stand against the might of military authorities to halt the exploitation of poor farmers (Akhtar 2006). Yet the fact remains that, despite achieving its immediate goal of halting evictions, the AMP has been unable to help landless tenants secure legal ownership rights to the land on the military farms.

The tenant farmers cultivating the military farms lack legal standing to prove their ownership of the military farm lands, despite the Chief Minister of Punjab's informal recognition of their rights in his talks with the AMP leadership in 2010 (Settle 2013). Just before the general elections, the current Prime Minister Nawaz Sharif held a rally in Okara, where he also promised AMP land rights. Whether the PML-N government will deliver on its election promises remains to be seen. Conversely, however, pressure from Pakistan's military–industrial complex continues to derail efforts to grant the landless tenants ownership rights to the military farm lands. The AMP leadership has also expressed fears that the military farm management is waiting for an opportune moment to reassert it control of the military farms and again evicting the AMP affiliated landless tenants off these lands (Karim 2014).

Assertions that landless tenants on military farms do not have to pay a major share of their crops to the military farms due to the AMP are not entirely accurate (Akhtar 2006; Mumtaz and Mumtaz 2012). The fact remains that tenancy arrangements had been changing in many of the villages even before the AMP became active. The AMP activists and villagers on the Tiryana Estate, for instance, pointed out that sharecropping arrangements had been altering in favour of lease arrangements long before the military farms tried to put in place the new contract system. One of the villagers identified Prime Minister Bhutto's attempts to improve tenant rights,[13] as the time when landowners, including the military farms, tried to discourage tenant farming in favour of leasing arrangements. The AMP activists estimated that while some sharecropping had existed in villages of the Tiryana Estate by the time their movement was launched, lease contracts had largely become the norm. According to these activists, their fight with the military establishment was not over switching from sharecropping to lease arrangements, but rather the military administrations attempt to increase lease rates. The AMP activists on the Tiryana Estate claimed that the military farm administration was demanding that lease rates be increased from Rs 26,000 to Rs 40,000, which is what their movement tried to resist, and in fact managed to achieve.

Due to the AMP resistance, tenants in villages on the Tiryana Estate were said to be paying a reduced lease amount of about Rs 15,000, which was less than what they used to pay when the AMP movement started. Yet the fact that farmers were still paying reduced lease rates discredits claims that the AMP had succeeded in stopping payment to the military farm establishment in Okara (Akhtar 2006).

When the AMP-led conflict took hold on military farms, including the Tiryana Estate, tenants were divided into 'white banner' and 'red banner' groups. According to the 'red banner' AMP activists, many more tenants were part of their movement initially, but when things became serious, a more conciliatory 'white banner group' of the AMP emerged which was willing to negotiate with the military farm establishment. The 'red banner' group, backed by the Labour Party, which had a less compromising stance with the military farm establishment, admitted fighting with the 'white banner group' of tenants, and even forcibly taking over their lands as the movement gained strength. The 'red banner' activists justified their actions due to the 'white banner' group's willingness to help the military establishment in discrediting the 'red banner' group as a '*qabza* group'.[14] This evidence of infighting between the tenants contradicts assertions, made by Akhtar (2006), for instance, concerning the cohesiveness of the AMP movement against a common adversary.

However, after the leasing issue had been settled, the 'white banner group' also returned to continue farming lands on the Tiryana estate. Moreover, 'outsiders'[15] were found to still be cultivating land on the military estates, which indicated that the trend of leasing to 'outsiders' had not been completely curbed either. Six outsiders in just one of the villages in the Tiryana Estate were cultivating between 25 acres and 50 acres of agricultural land.

The AMP activists in the Tiryana Estate admitted that the military farm establishment was still creating problems for the AMP-backed tenants, such as withholding water from tube-wells to supplement irrigational requirements especially for water-intensive crops such as rice. While these problems are noted on the Tiryana Estate in particular, this military farm was considered to be at the epicentre of the AMP struggle, and the emergent problems now visible here do undermine broader claims concerning the AMP's success in solving tenant problems.

Since wage labourers were also embroiled in the conflict, there was an expectation that the eventual conferring of ownership rights would enable them to gain the benefits of both agricultural and homestead land. There was, however, limited evidence to this effect, at least in the case of the Tiryana Estate. After a particularly violent spate, which saw three AMP activists lose their lives on the Tiryana Estate, AMP activists marched to register their grievances with the chief minister, who made the decision to provide them with 5,386 residential plots of 10 *marlas* (0. 5 acres) each for landless tenants on the Tiryana Estate. While a few widows and other poor households got land free of charge, most beneficiaries had to pay between Rs 5,000 and Rs 6,000 to get this land. It is important to note that this fee was being paid to the AMP leadership instead of the government. Why the AMP was charging this amount is discussed in more detail below, with reference to the role of NGOs/INGOs in supporting the AMP.

Agricultural labourers were, on average, said to be getting higher wages now that the AMP had exerted itself against the military farm management than what they were being offered by lease operators who had been brought in by the military farm administrators from outside the village. In villages in the Haryana Estate, their wages were reported to have gone up to around Rs 200 from between Rs 120 to Rs 150 per day. However, further exploration of the wage increase revealed that this amount had gone up over a considerable period of time, and that these wages are not much higher than what agricultural labourers are paid by landowners in the surrounding farming areas of the district. Hence, participation in the AMP movement did not prove to have significantly improved the socio-economic position of agricultural labourers working in villages of the military farm where I conducted field research.

b. Increasing fragmentation and myopia in the AMP

After its initial success, fragmentation within the AMP has been growing, a visible indication of which is the increasing marginalization of women within the movement more broadly, as well as growing tensions between the male and female leadership. While the tensions between the male and female leadership pertain to personal grudges concerning sharing power and other benefits associated with leading the AMP, there is also evidence of a narrowing vision among the leaders of the movement, which is further undermining the potential of the AMP to challenge the status quo.

Women in the Indian subcontinent have a long history of being mobilized to take part in various resistance movements, including the struggle for independence from British colonial rule. However, the basic question to ask here is whether female participation in resistance movements leads to any positive outcomes with regard to gender relations at the household level, or even in women's own sense of empowerment. More often than not, and irrespective of the outcomes of the broader movements themselves, patriarchal patterns have been observed to revert as women go back to their accepted socially defined roles when the most arduous phase of resistance subsides (Mumtaz and Mumtaz 2012). Similar problems have occurred in the case of the AMP movement as well. Affiliation with the AMP movement initially provided women the opportunity to gain more confidence, exposure, and mobility (Mumtaz and Mumtaz 2012; Toor 2011). Yet, after the AMP movement became more established and achieved increasing recognition, the women affiliated with it continued reflecting patriarchal values which subordinate women's place within the social order in rural Pakistan. While women participated in the struggle for land rights of tenants on the military farms alongside men, many of these same women still thought that the right of landownership belonged to men. Besides the above-cited research, women activists I spoke with in villages located on Tiryana Estate also confirmed the prevalence of such perceptions. Hardly any of the women involved in the movement had any agricultural land allocated to them individually.[16]

Gender-related problems have also emerged at the leadership level within the movement. Women did secure leadership roles within the AMP at the village and district levels, especially due to the insistence of NGOs, which have worked with the movement. Some of the women leaders also got a chance to represent the AMP on national and international platforms. For example, Anila (not her real name), the president of another district's wing of the AMP, along with its finance secretary, participated in the World Social Forum (WSF)[17] in Mumbai in 2004 and in Karachi in 2006. However, a Lahore-based head of an NGO for women, which had been working in solidarity with the women in the AMP, pointed out how the level of female empowerment in the leadership was inconsistent across the different geographic areas in which the AMP was active. Other research has also noted women activists in the AMP complaining about the lack of information sharing in regard to the AMP budgets and expenditure, and major decisions being made in the movement, despite their names being included as female leaders of the movement (Mumtaz and Mumtaz 2012).

I spent a considerable amount of time with the Tiryana Estate Female Wing, President Bushra (not her real name),[18] in order to explore the issues of gender-related and other tensions, which have emerged in the AMP. Bushra was in her mid-forties and had been educated to secondary school level. Her husband worked in a private security firm in Lahore, as a security guard. One of her brothers was a *patwari* in the same district. Bushra's family was not directly involved in tenancy, yet she became involved in the movement due to her earlier social welfare work when she was living with her husband in Pakistan-administered Kashmir. Bushra mentioned that she had been invited by the AMP Male Wing President for Tiryana Estate to take over the position of the Female Wing President, due to her relatively higher level of education and exposure as compared to other local women. While not a peasant woman herself, Bushra did well in a leadership role in the AMP. She pointed out, for example, how she had helped bring over 1,000 women from the surrounding villages to a recent peasant gathering. She was also liked by other villagers and AMP activists.

Yet, despite her commitment to the movement, and her ability to assume a leadership role, Bushra admitted she was experiencing growing tensions with the President of the Male Wing of the AMP in the Tiryana Estate. Bushra described how the Male Wing President, at the behest of NGOs and the Labour Party, which was supporting the movement in the Tiryana Estate, had brought her in to fill the position of the Female Wing President, but he did not really want to share power with her. Often, he would not even inform her that AMP meetings were taking place, and instead take other male members formally and informally affiliated with the AMP. Bushra accused the Male Wing President of not only being reluctant to share power with her, but also of trying to keep her at bay so that she would be unable to see him squander AMP funds.

Influenced by civil society organizations, which had been trying to provide advice to the AMP, its leaders had realized the value of trying to generate funds to achieve self-sustainability. In other areas, AMP fund raising may be subject to variation. In the Tiryana Estate, however, funds were being raised by charging Rs 1,000 for each *acre* that the AMP had managed to negotiate and return to tenants. Residential plots allotted by the Chief Minister to compensate villagers who had been subjected to violence, were free of cost, but the AMP was charging landless tenants Rs 2,000 to allocate them one of these ten *marla* plots. During a visit to her village in mid 2011, Bushra informed me that she herself had raised Rs 1,600,000 for the AMP just from her own village through the above-described fees. The AMP activists in some of the other village of the Tiryana Estate also mentioned collecting money from their villages. Bushra and other AMP activists pointed out that the money raised by them was used to fund AMP activities, such as attending meetings to garner further support,[19] as well as to meet the costs of litigation filed against AMP activists who had been arrested after clashing with the military farm establishment, and with 'outsiders' being brought in on lease arrangements to cultivate land from which Tiryana Estate villagers had been evicted.[20]

Several times Bushra complained that she had not been adequately rewarded for her contributions to the AMP. However, I later learned that Bushra had also been given 6 acres of land where her family now grew rice. She confirmed that this was so, but then pointed out to me that the land allocated to her was much less than that allocated to the Male Wing President. She, however, did not acknowledge the fact that her family was previously not tilling land, and therefore they were not entitled to land resumed by the AMP for the purpose of being returned to previous tenants. Instead, she mentioned how her husband often chided her about her inadequate remuneration in light of all the efforts she had made, including the beating she had suffered,[21] for the AMP.

Bushra complained that the AMP's Male Wing President for the Tiryana Estate was barely educated, and that he did not believe in educating his daughters, or girls within their community either.[22] She further asserted that the male leadership of the movement, including the Male Wing President, primarily favoured using coercive tactics to continue their struggle. This would entail focusing on demonstrations, and even resorting to further violence to deter outsiders from leasing land in their villages. Bushra considered these measures necessary when strong resistance against the military farm establishment was needed, but now she felt the movement needed to think of adopting other strategies. Bushra claimed that while women in her area understood this basic fact, the Male Wing President and other male activists close to him did not realize this.

Thus, besides her personal animosity, Bushra's criticism did raise a broader concern regarding the capacity of this grass-roots resistance movement to sustain itself on the basis of coercive strategies alone. Bushra herself emphasized the need for focusing on education, which she considered vital for achieving gender empowerment within her community. She pointed out that there was only a primary school in their village, and many parents were reluctant, or could not afford, to send their daughters for secondary schooling outside the village.[23] In view of this fact, Bushra's own daughter had begun providing home schooling to nearly a dozen girls who had dropped out after primary school. Bushra was keen to register her efforts as an independent NGO, to be able to open a high school for girls within her community, as well as to distance herself from the AMP. Due to tensions within the AMP, the male leadership was not being very supportive of Bushra's efforts.

My discussions with the male AMP activists revealed that the district administration had recently acknowledged tenant rights to stay on the military farms, yet the legal ownership issue of the military farm lands had not been resolved. These activists realized the value of obtaining legal status for the lands their families had been cultivating for generations. This issue was thus on the list of priorities for the male AMP leadership on the Tiryana Estate. However, beyond the need to help activists who had violently resisted evictions, and to continue to safeguard against future threats of eviction, there was not much evidence of a coherent strategy for securing land rights from the military farms or working with the AMP in Tiryana Estate to achieve other development goals.

Initially, the AMP provided a cohesive platform of resistance to all landless tenants irrespective of existing social divides, and it even enlisted women in support of its cause. However, the problematic treatment of women in the movement has now become a serious problem undermining the initial sense of solidarity. Besides fragmentation along the lines of gender, the dominant male leadership also had a narrowing vision, which is preventing the AMP from transforming itself from a reactive mobilization movement into a proactive social platform capable of achieving meaningful change.

c. The contested role of NGOs/INGOs in supporting the AMP

With little recourse to securing support from existing state institutions or mainstream donor mechanisms, the AMP sought and received support from local and international NGOs and leftist parties within the country.

Before examining the specifics of civil society inability to lend effective support to the AMP, let us briefly take note of the larger role played by different civil society organizations in addressing the concerns of poor farmers. Earlier chapters have already discussed efforts by QUANGOs like the Rural Support Networks that employ market-based strategies like micro-finance to address problems facing poor farmers. But there are also examples of other rights-based international and national NGOs working to enable transformative socio-political change to bring about improvements in the lives of poor farmers. Their programmatic interventions aim to uplift marginalized and neglected poor farmers by not only providing them with inputs or building their capacity to increase productivity, but also helping organize farmers to collectively influence policies and practices directly effecting their lives.

In Pakistan, national and international NGOs have also shown the capacity to initiate multiple responses to important issues concerning poor farmers, ranging from direct programmatic interventions to initiating research and formulating advocacy campaigns. They have also served to link local activism by marginalized farming communities within broader development policy discourses. For instance, national and international NGOs have helped organize *hari* (landless farmer) conferences, which have been drawing together thousands of male and female poor farmers and agricultural workers. During such events, demands have been made for amendments to tenancy laws to protect landless farmers from exploitation by large landowners, and to offer better terms and rights to agricultural labourers. NGOs have, however, problems ensuring the consistency and cohesion of collective action. Several NGO personnel themselves admitted that *hari* conferences, for instance, are held infrequently, and remain dependent on funding from external sources.

Bano (2012) has identified a detrimental role of foreign aid on community participation arguing that aid undermines existing levels of community mobilization. Within this context, Bano discusses the case of the AMP, to understand why this particular indigenous movement succeeded in undertaking successful collective action against the powerful military. According to Bano, the AMP was able to act as a cohesive movement by foregoing foreign aid and instead developing close ties with local Marxists (belonging to the now defunct Communist Party and the Labour Party), as well as the People's Rights Movement (PRM). Bano particularly highlights AMP's linkage with the PRM, a miniscule entity (made up of six core members), which aims to provide a platform for local indigenous movements, including fisherfolk and slum dwellers movements, as well as the AMP. Bano credits the PRM for helping sustain the AMP and convincing it not to rely on aid from international development agencies (with the exception of receiving support from ActionAid, although this arrangement is also described as having been severed subsequently by the AMP). Conversely, it is the PRM which is credited with linking the AMP with international media and bringing them to the attention of foreign embassies, whereby strategically boosting the AMP's visibility and enabling it to take an assertive stance to counter state repression in the glare of public scrutiny.

While PRM may have provided valuable support to the AMP by helping plan an active media campaign and lobbying on its behalf with legislators and the international community, PRM was certainly not the primary entity working to support the AMP. Bano's analysis of the situation seems informed by very initial contacts with the movement. By the time I went to conduct fieldwork in Okara, the movement had been in existence for over eight years and there were numerous stakeholders involved in supporting different factions of the AMP. In fact, a range of national and international NGOs and other leftist groups (including the Labour Party of Pakistan), have not only tried to support the AMP, but also exerted considerable influence over the movement. This support came in the form of trying to lend the movement organizational support, provide strategic advice, and undertaking capacity building of its leadership. However, the varied forms of support provided to the AMP, including that by the PRM, has remained inconsistent and proved to be short-sighted, as a whole.

There is a precedent for peasant movements becoming fragmented due to competition for funding, fights for public recognition and fame, and personality clashes (Borras *et al.* 2008; Borras 2009). Perhaps if the INGOs and leftists parties in Pakistan had paid heed to lessons emerging from the experiences of international peasant movements, they could have been able to pre-empt such problems from afflicting the AMP. They were, however, unable to avert the fragmentation of the AMP, or to help broaden its vision and scope of activities beyond confrontation with the military establishment.

In 2005, there was an evident split in the AMP. On one side was the Okara group, and on the other the Khanewal group. Some of the political analysts and activists I spoke with regarding the AMP blamed this split on the difference of approaches between the leftist political parties, which supported one group of the movement, and INGOs, which supported the other group. This split apparently occurred when different civil society organizations began working with different segments of the AMP, began emphasizing different issues, and even imposed different agendas on the AMP movement. A leftist political party, for instance, lent support to the Tiryana Estate male leadership in helping resist the military farm administration, but it did not pay much attention to gender issues. Other NGOs, which chose to lend support to the AMP movement, especially INGOs, were motivated by the imperative of helping villages where religious minorities were present.

Earlier research has noted how the AMP leadership was drawn from both Christian and Muslim communities, and it went on to claim that there was unity among the ranks since land rights were a common cause for these different religious communities (Mumtaz and Mumtaz 2012). While Akhtar (2006) conceded some latent tensions, particularly between the Muslim majority and the Christian minority communities, to still be apparent in daily life, he did not view these as deterrents in terms of the overall cohesion of the AMP. However, some of the NGO representatives I contacted, who had worked with the AMP movement since the time of its creation, pointed to much more serious differences within the AMP movement emerging along religious lines, which ultimately exacerbated the splitting up of the movement between the Muslim dominated Okara group and the Khanewal group, in which minority (Christian) communities were also influential.

Some of the political activists I spoke with blamed donor-driven NGOs for trying to superimpose their own organizational imperatives onto the AMP, which disrupted its organic evolution and capacity for bringing about transformative change. For instance, a Lahore-based activist blamed NGOs for having imposed their own bureaucratic hierarchies onto this social movement, with an insistence of demarcating leadership cadres, which had caused friction and infighting in the indigenous movement, undermining its initial social solidarity. The activist further asserted that the 'hegemony' of the NGO sector in Pakistan itself needed to be questioned, which could only be done if NGOs became more accountable to the causes they aimed to serve, rather than the donors who provided them funding.

Another women's rights activist claimed that most NGOs were unable to prevent the marginalization of women in the AMP, based on the assertion that no serious effort was made to give women land rights. Instead of providing international media exposure, and linking the AMP movement with other peasant movements around the world, the activist blamed NGOs for picking up a handful of AMP representatives to in turn publicize their own organizations, by taking them around to conferences in major cities, held in expensive hotels, while ignoring the majority of poor farmers affiliated with the AMP. Such criticism was not unique, and finds resonance, for instance, with critiques presented by others, including Tariq Ali (2006), who claims that NGOs are stifling genuine social movements.

Furthermore, despite sponsoring the participation of selected AMP leadership to participate on platforms such as the World Social Forum, major INGOs also remained unable to help create effective linkages between the AMP and other peasant movements. AMP was not effectively linked to Via Campesina, for instance, which describes itself as an international movement which coordinates peasant organizations of small and middle-scale producers, agricultural workers, rural women, and indigenous communities from Asia, Africa, America, and Europe. Via Campesina had mentioned the AMP in one of its publications highlighting a dozen peasant movements around the world in November 2003 (Via Campesina 2004). This mention was, however, brief and simplistic, and merely highlighted state-led violations against AMP members. I found no other linkages between Via Campesina or any other international peasant movements and AMP activists in the Tiryana Estate in particular. Based on these observations, it seems reasonable to conclude that the NGOs and other leftist groups working with the AMP do not seem to have devoted sufficient time for alliance-building activities in order to help the AMP extend the reach of their collective actions beyond their own ranks. Sponsoring a representative of a peasant movement to attend a conference, or providing enough information to help publish a profile on a local resistance movement by a transnational peasant movement, is not the same thing as creating effective synergies between a local movement and broader international coalitions which can help guide, sustain, and learn from a Pakistani peasant movement.

Thus far, civil society efforts remain unable to support and sustain social movements which can work towards long-term social change in the country, that in turn may challenge the underlying causes of structural inequalities perpetuated through elite-led and market dominated agricultural development strategies. Civil society in Pakistan has not yet succeeded in developing a comprehensive strategy which can be employed to disable the coercive apparatus of the state, to obtain access to political power, and create the conditions that would give rise to a consensual society wherein no individual or group (in this case poor farmers) is reduced to a subaltern status.

d. Contextualizing AMP's achievements

The AMP provides a clear-cut instance of a salient, and very recent, example of social mobilization from below. Such mobilization of poor farmers remains of vital significance in terms of bringing about meaningful change (Borras 2008). However, the movement lacked support from either statist reform initiatives, or sufficient support from a wider range of civil society organizations, to be able to bring about revolutionary change.

The resistance put forward by the AMP also cannot be categorized as a revolutionary form of violence. Ahmad (2007) points to various forms of violence prevalent in Pakistani society, ranging from domestic, criminal, official, and ethnic, to religio-sectarian and terrorist. However, all these forms of violence are differentiated from the revolutionary form of violence, since it is only the latter which seeks systemic change. It is the lack of the revolutionary form of violence, as described by Ahmad, which explains the persistence of elitist and colonial cultures in the country.

While the AMP used violence in response to violence perpetrated by the state itself, the AMP's reactive use of violence lacked the proportional capacity to pose a serious threat to the state. Conversely, the military establishment itself is well entrenched in the institutional set-up of the state, and it was able to bring in other state institutions such as the police and Rangers to help repress the AMP. The military has now also implicated AMP activists who restored to counter-violence in criminal proceedings. Yet, AMP's use of violence has lacked transformative capacity, since it has been unable to secure strong support from other prominent political or social movements.

Due to resorting to the use of violence to counter state repression, the AMP instead experienced a withdrawal of support for its cause. A senior officer of the premier human rights-based NGO in Pakistan mentioned how they had launched a campaign to highlight the violent repression of the AMP by the military farms administration. However, the human rights-based entity began feeling increasing discomfort when the AMP itself began using violence as a tactic to resist evictions. Mohammad Hassan (not real name), a prominent political figure who served in a senior position under the Bhutto government when the land reforms for 1972 were introduced, made similar comments. Mr Hassan mentioned that leftist political parties had consulted him while trying to give coherence to the AMP movement, but they did not heed his advice concerning the need for a softness/flexibility of approach, and when the AMP adopted the hard-line slogan '*malkiyat yah moat*',[24] Mr Hassan claimed he began to distance himself from the movement.

The AMP has been categorized, from within the movement, as well as by outsiders, in divergent ways. Depending on the stakeholders involved, the AMP has been described as an indigenous mobilization of poor farmers seeking social justice, or, conversely, as a potential rural constituency for aspiring leftist groups with political aspirations, and even as a 'criminal organization'.[25]

Nonetheless, the AMP leadership itself remains fragmented and insecure about its victory, fearing that the military land being cultivated by the landless tenants is in the process of being allotted to multinational agri-business companies for corporate farming (Karim 2014). Despite repeated declarations by several politicians of major political parties, proprietary rights for the Okara tenants had not been granted until early 2014. Part of the problem is that, if concessions are made for tenants in Okara, a problematic precedent would be set for making similar concessions for hundreds of thousands of other landless farmers cultivating state-owned land leased to different government departments, including the Ministry of Defence, across other parts of the province. Had this precedent been set, the AMP could have rightly claimed making a major dent in the prevailing 'political economy of defence' in the country (Jalal 1995: 140). As things stand, however, such a claim is not tenable.

Some of the emergent literature, and even media coverage of the movement, however, seems to have overestimated the AMP's significance (Akhtar 2006; Bano 2012; Toor 2011; Zaidi 2012b). However, despite the above evidence, which does dilute some of the more enthusiastic claims concerning the AMP's accomplishments, it is nonetheless clear that this was a spontaneous and indigenous resistance movement, which took a very strong and motivated stance against the most powerful state institution in the country, the military. A range of factors, including internal tensions within the AMP movement, as well as the incapacity of external entities to help overcome these internal frictions, have created ruptures within the movement which prevent it from transforming itself into a broader social movement representing poor farmers around the country. Whether the AMP leadership, and civil society organizations supporting it, will realize their shortcomings, and initiate alternative strategies to reinvigorate the movement enough to be able to alter the existing landownership patterns in rural Pakistan, remains to be seen.

Conclusions

The fact that the AMP movement was catalyzed by the military's attempt to alter traditional land tenure arrangements into contract-based farming is indicative of the fact that the military itself has also been influenced by the increasing commercialization of land markets. Like the landed elite, the military, too, has the ability to draw support from other state institutions to maintain its control over agricultural land. However, the very fact that the AMP resisted the military, and managed to halt evictions, demonstrates that poor and landless farmers do have some form of agency or capacity to act independently and to resist attempts to exploit them. Yet, the prevailing structures of power serve to limit the choices and opportunities available to poor farmers to achieve transformative change, such as preserving land tenure arrangements from increasing commercialization, or obtaining legal titles of land they have been cultivating for generations.

While the situation for farmers facing the threat of eviction from military farms has visibly been halted due to the resistance put up by the AMP, the victory for the movement is not as complete as earlier claimed, nor is the movement as cohesive as some of the recent literature on it suggests. Furthermore, since the AMP has been unable to secure legal landownership rights of landless tenants working on military controlled farms, its success remains partial and potentially reversible.

There are evident tensions between the leadership and the farming communities involved in the AMP along gender and religious lines, which have begun to undermine the initial solidarity of the movement. Despite civil society support for its cause, the AMP was not provided effective guidance or support to sustain the initial sense of solidarity which helped spark the movement.

Given the dearth of state-led or donor efforts aiming to alter the existing inequitable distribution of resources, including the lack of any legislative efforts to undertake redistributive land reforms, or provide more secure rights for tenant farmers, social mobilization or even resistance from the grass-roots level lacks recognition and legitimacy among policy makers. Instead, the state and international donors continue to rely on the landed elite, in using market-based strategies for achieving agricultural growth. Given this convergent nexus, there is little room for mobilization from below, as symbolized by the AMP, to challenge the vested interests of an institution as powerful as the military.

Notes

1 The Anjuman Mazahreen Punjab translates from Urdu into English as the Association of Landless Peasants of the Punjab.
2 Dullah waged war against the Mughal Emperor, Akbar, whose reign extended from 1554 to 1605.
3 Sufis were Muslim mystics, many of whom played an important role in the spread of Islam in the Indian subcontinent, and they subsequently sympathized with, and even led, peasant movements motivated by notions of humanism, and in rejection of the exploitation of tillers of the land by the Mughal Empire and its administrative machinery. *Sheikh* is a term which refers to Sufi leaders of a given school of thought.
4 Some of their specific demands included a call for the cancellation of peasant debts, and the exemption of unjustified and un-economic holdings from taxation, by the British colonial government.
5 The English translation of the Urdu name Mazdoor Kissan Party would be Labourer and Farmer Party.
6 NWFP was renamed Khyber Pahktunwah Province (KPK) in 2010.
7 The PFF movement was catalyzed by the increasing commercialization of the fishing sector, whereby the government began auctioning several fresh water bodies, depriving indigenous fishermen families of the right to fish in these waterways. Staunch resistance by the PFF compelled the Provincial Assembly of Sindh to pass a bill abolishing the contract system through its Sindh Fisheries (Amendment) Bill 2011. Despite this new legislation, however, fishermen are still complaining of their inability to access water bodies, particularly in areas where influential landowners have a strong presence.
8 The AMP eventually spread to nine other districts, namely: Multan, Khanewal, Jhang, Sargodha, Pakpattan, Sahiwal, Vihari, Faisalabad, and Lahore.
9 Lower Bari Doab Canal (LBDC) comes out of the Balloki Barrage, which is located 65 kilometres south-west of Lahore, the provincial capital of the province of Punjab.
10 The Pakistan Rangers are part of a paramilitary force tasked with maintaining security in areas of conflict within the country, besides being called in to supplement war efforts. The Rangers recruits are army personnel, but they are administratively controlled by the Ministry of Interior, which also controls the police.
11 Besides different castes and kinship groups, there were also Christian tenants on the military farms. Christians in the Punjab were typically converts from the lowest untouchable castes, and there was some provision under the British canal colonization schemes to provide land to the 'depressed' classes in which category Christians were accommodated.
12 Marxist theories of peasant revolt have identified different strata within the peasantry that adopt widely varying roles in rural conflict. Marxists analysis of farmer revolts in the Indian subcontinent, for instance, indicates a leading role for the middle peasantry (commonly understood as self-cultivating farmers who also own land but typically do not exploit the labour–power of others, except that of their

own household members) in creating impetus for a revolution. But this middle peasantry was also criticized because of their limited social perspective due to their class position, which often led them to move away from revolutionary movement (Alavi 1973).

13 Land reform legislation (as in the Land Reform Act of 1972) tried to regulate sharecropping rents, typically preventing the landlord from obtaining more than 40 per cent of gross production, in addition to reiterating other legal rights of tenants (Herring 1983). While these measures were not enforced effectively, they did instigate the trend of landowners taking land back from sharecroppers to avoid legal complications, and begin leasing it out on a contract basis instead, which provided upfront rents instead of taking a share of the crops grown by the sharecropping tenants.

14 Literally the term means land-grabbing group. Land grabbing by criminal elements using force to encroach on land remains a common phenomenon in Pakistan. The army's attempt to label the AMP as a '*Qabza*' group was meant to discredit them and the underlying imperatives of their movement.

15 The term 'outsiders' refers to contract farmers from outside the villages who were being brought in by the military farm to replace local tenants, the growing presence of whom in the military farm villages had sparked the AMP resistance movement in the first place.

16 There are exceptions, however, given that some of the women activists who held leadership positions within the AMP did get some land for services rendered to the movement. This issue is, however, explored in more detail in the subsequent discussion.

17 The WSF is an annual meeting of civil society organizations from around the world seeking to highlight alternative pathways to development and empowerment of the global poor.

18 The AMP movement has been structured differently according to which entities support them. In Okara, where the Labour Party has been involved, the AMP leadership cadres have been organized at the village level and at the larger military estate level.

19 Bushra suspected the male leadership was siphoning, or at least wasting, funds obtained from the tenants by attending such meetings, without accounting for the expenses incurred to attend the events, or disclosing how much money had been gathered during these fundraising activities.

20 Some of the legal cases lodged against the AMP leadership were very serious, including murder charges, since some of the 'outside settlers' also died in the ensuing violence on the military farms. The AMP leadership however denied these charges and claimed it was the military leadership which had implicated them in false cases, and that their activists had not killed anyone. The veracity of these charges is an issue which I could not investigate further.

21 This statement referred to the above-mentioned instance of an AMP rally being baton charged, and in the chaos, Bushra was hit in the head and had to get stitches.

22 Bushra made these claims when we were alone rather than in the company of other AMP activists. Moreover, while I met several other male activists affiliated with the AMP in the Tiryana Estate, I was unable to interview the Male Wing President since he was always away from the village during my visits.

23 Parents of girls going to a secondary school located several kilometres away had to hire a motorcycle rickshaw to transport their girls to and from the school.

24 Literally the Urdu phrase means 'ownership or death'.

25 Military farm managers in particular have tried to discredit the AMP by claiming that it is this movement which is preventing tenant farmers from entering into well-remunerated contract system farming with the farm management, since its leadership is interested in pilfering surplus from the military farms in connivance with dishonest sharecroppers.

7 Conclusion

An attempt has been made in the preceding chapters to provide a comprehensive approach to understanding why poor farmers continue to remain marginalized in developing countries like Pakistan today. In doing so, I have pointed to the vital need for understanding the combined interaction of a broad range of state and donor interventions which have significant implications for poor farmers. Despite optimistic claims surrounding peasant resistance movements, an effort was also undertaken to point out several problems afflicting resistance attempts on the ground, undermining the attempt to achieve meaningful change.

Extensive field research was undertaken to identify relevant state institutions and policies, and how they interact with particular donor-supported policy and programmatic interventions across rural areas of Pakistan. Field sites selection was itself guided by the need to obtain particular data needed to address each of the three main research questions. Similarly, the research respondents were chosen from a wide range of stakeholders, including varied levels of government decision makers, donor agency, and civil society representatives, and different categories of farmers, to solicit a broad range of perspectives concerning the main research questions.

The resulting research analysis presented in this book confirms that there are myriad ways in which state and donor agencies have an impact on the lives of poor farmers. Besides targeted state and donor programmatic interventions aiming to specifically address poor farmers' concerns, a much broader set of state institutions and donor policies exert influences on poor farmers in unexpected, and therefore often neglected, ways. The vested interests of the landed rural elite are preserved through wide-ranging state support, including political patronage as well as bureaucratic procedures that are not even directly involved in the process of agricultural development or poverty alleviation in rural areas.

Generally, policy and decision makers in Pakistan have failed to adequately protect poor farmers from varied levels of exploitation and deprivation. Subsequent to lacklustre initial efforts to undertake land reform, ruling political parties have failed to bring land reforms back onto the political agenda. The urban-based MQM prepared a recent proposal for the redistribution of land, but its formulation was viewed as a self-serving move to undermine the rural support of its political opponents, rather than a genuine effort to bring the issue of land reforms back onto mainstream political agendas. Institutions such as the judiciary have also not played a proactive role in altering the inequitable landownership situation in the country. The Supreme Court, in fact, declared land reforms to be 'un-Islamic', a decision that has not yet been revoked, despite recent public interest litigation calling for a review of the longstanding anti-reform judgment. Unless the ownership and use of land is regulated to address the prevailing disparities, an exploitative configuration of forces will remain in place which establishes a unidirectional flow of agricultural production benefits from poor farmers to the larger landowners.

Donor agencies have not paid much attention to the need for redistributive reforms in Pakistan either. Instead, donor pressure has resulted in market-based policies being endorsed through legislative attempts. The Corporate Farming Ordinance 2001 and the Plant Breeders' Rights Bill 2011 provide an impetus for the corporatization of agriculture and the protection of patent rights for transnational agri-business concerns. Such legislative efforts can provide opportunities for larger landowners to increase yields by investing in more expensive inputs, or by leasing out their lands to corporate farms, but they imply adverse effects for poorer farmers by making inputs more expensive and compounding land scarcity.

No serious efforts have been taken in Pakistan by legislators or government officials to protect the interests of agricultural workers or women involved in agriculture. There is an absence of effective legislation protecting their rights, by requiring, for example, landowners to enter into formal contracts specifying work requirements or wages for agricultural labourers. After coming to power in 2008, the democratically elected PPP government took some measures to facilitate women's ownership of land. The Anti-Women Practices Bill 2011, however, remains problematic due to its myopic and primarily punitive approach towards addressing the problem. This bill ignores both the socio-cultural compulsions, which prevent women from asserting their land rights, as well supplemental institutional measures needed for ensuring that they in effect can secure their rights.

Adopting a less radical approach than redistribution of land owned by the rural landed elite, the Benazir Landless *Hari* Scheme recently tried to give some state-owned land to a limited number of landless farmers across rural Sindh. While this programme recognized the need for effective targeting, and providing landless farmers access to a range of other facilities besides land, the actual implementation of this scheme encountered several difficulties. Even the NGOs involved by the government to help implement this scheme have been unable to ensure that only deserving landless farmers received the state land grants.

The prevailing dearth of institutional support is compounded by pressure from agencies like the World Bank to curb public expenditures, while raising more revenues, which has led the state to devise policies which place an equal burden of responsibility on large landowners and poor farmers alike. For instance, the government has implemented flat-rate charges in the irrigation sector, which puts a uniform burden on all farmers, rather than trying to make the state taxation mechanism more progressive by charging differentiated rates for poor farmers and larger landowners. The lacklustre approach by state institutions towards implementing the Agricultural Income Tax (AIT) has resulted in the AIT continuing to charge a meagre amount on land-ownership, instead of securing incremental revenues based on actual agricultural incomes. Such measures demonstrate the state's inability to place greater pressure on larger landholders. The large landowners' ability to resist taking a great share of the revenue generation burden, in turn, has negative impacts on poorer farmers, who experience an ever-diminishing capacity of public sector institutions to improve the efficiency and capacity of vital infrastructure, such as irrigation networks. Given the prevailing institutional biases against the poor, larger landowners continue to monopolize the scarce state resources, which are available, while poor farmers continue to experience problems like water shortages.

Ironically, the Pakistani state safeguards the interests of the landed rural elite, even if some of the rhetoric employed to preserve these interests is that of trying to help poor farmers. The ongoing subsidization of fertilizers and pesticides, as well as government control over the procurement of wheat, indicate how the state resists donor pressure to allow market forces and the private sector to provide all agricultural inputs and commodities. Instead of helping poor farmers, however, there is evidence of massive corruption and ineffective targeting of such subsidy schemes. In the case of wheat procurement as well, it is larger landowners and commission agents who derive benefit from government control prices rather than poorer farmers. Donor agencies recognize that larger landowners and other intermediaries are capturing existing government subsidization schemes, but they have not been able to put in place effective programmes either, which could ensure that poor farmers become the primary beneficiaries of these schemes.

There is little evidence of donor support for programmatic interventions focused on assisting poor farmers in connecting with opportunities presented by the market, or to buffer them from the exploitation induced by globalized market forces. In strategic frameworks like the Poverty Reduction Strategy Papers, prepared by developing countries like Pakistan to access IMF and World Bank concessional loans, the use of growth-led policies continues to be emphasized to achieve not only agricultural growth, but also rural poverty alleviation. Both the PRSP 1 and 2 for Pakistan stress the need for increasing productivity and value addition in agriculture, and the adoption of new technologies, even though these are not relevant policies for poor farmers. Despite euphemistic claims of providing the poor a chance to participate in the process of their own development, the PRSPs formulated for Pakistan failed to identify concrete measures to improve the lives of landless farmers, agricultural labourers, or women involved in agriculture. There is also a lack of attention paid to overcoming particular barriers faced by poor farmers in terms of assessing export markets, nor have any programmatic interventions been formulated to remove distortions in agricultural labour and capital markets, which tend to disadvantage poorer farmers in particular.

On the other hand, donor agencies like the World Bank are trying to use NGOs to encourage poor farmers to be integrated into capital and land markets. This donor-endorsed approach of using NGOs to allow poor farmers access to needed capital was tested in rural Sindh, where prominent QUANGOs/ NGOs like the RSPs have ongoing micro-lending programmes. Research conducted in villages in Badin district in particular found poorer farmers becoming more indebted due to the need to service interest rates of micro-credit provided to them, rather than being empowered by such efforts. Yet despite the evident difficulties encountered by attempts to use intermediaries to integrate the rural poor into capital markets, endorsement of the market system continues.

The World Bank has begun supporting an initiative to create more efficient land markets to address the problem of rural poverty. This effort aims to make land record management more efficient by computerizing land records in the Punjab, through the Land Record Management Information System (LRMIS) programme. LRMIS claims to have explicit poverty alleviation goals as well, such as lessening the burden of litigation and preventing the abuse of poor farmers by corrupt land revenue officials. Yet the programme is, in effect, focused on digitizing existing land records rather than correcting them, or using the land records to undertake any redistributive scheme. Drawing on experiences emerging from similar programmes of land record computerization in India, the ability of the LRMIS to produce better outcomes for poorer farmers seems problematic. It is, therefore argued that the LRMIS is poised to facilitate expedient land transactions, which, in turn, will help large landowners lease lands to capital-intensive agri-businesses.

It is also interesting to note significant complementarities within donor policies themselves. The computerization of land records, for example, provides the impetus to further liberalize land markets and enable agri-business investments in Pakistan. While the World Bank itself encourages corporate farming in order to achieve agricultural development and create more rural employment opportunities, the experiences emerging from a leading corporate farm in Pakistan are not encouraging, particularly for poor farmers. The Tarakee Farm is considered to be the most innovative, efficient, and profitable corporate farm in the country, but its remunerative benefits remain limited to the senior management. The Tarakee Farm has worsened land availability for sharecropping, and it is also driving up lease rents for those who cannot afford to pay advance payments for lease arrangements over multiple years. The Tarakee Farm itself does not offer sharecropping opportunities to poorer farmers, nor does it provide adequate remuneration to daily wage labourers or contract workers employed by it. Yet, an USAID development project aiming to enhance Pakistani farmers' access to international markets, and to create jobs and stimulate the economy, is also found to be collaborating with this capital-intensive corporate farm to export its mangoes to European markets, without ensuring that the resulting profits are shared by the seasonal agricultural workers employed by this farm. Therefore, while corporate farming may spur agricultural growth and profit margins for large entrepreneurial farmers, this business model seems to offer little prospect for helping poor farmers.

The World Bank's attempts to 'modernize' Pakistani agriculture by promoting market-based strategies such as the creation of land markets, or encouraging agri-businesses to address rural poverty without addressing underlying causes of deprivation resulting from unequal landownership patterns, therefore, seem problematic. Liberalization, as it is implemented within a developing country like Pakistan, reinforces the vested interests of the landed rural elite, without adequately empowering poor farmers to secure required institutional support from the state, or to overcome distortions of the market mechanisms which work to their disadvantage.

Instead of confining the analysis in this book to specific state or donor interventions alone, the broader interaction between state and donor attempts is highlighted for its conjoined inability to transcend elite-led agricultural development policies despite rhetorical assertions of benefiting poor farmers. On the one hand, the very nature of state formation in Pakistan helps us understand how and why the vested interests of the landed rural elite continue being preserved and propagated. The emphasis on using the market mechanism by the World Bank to further draw developing countries into the broader global production system also offers limited opportunities for poor farmers who lack the resources to invest in capital-intensive cultivation, and have limited direct access to markets. Instead, it is large landowners who have the resources needed to engage in capital-intensive corporate farming, and to purchase increasingly expensive agri-inputs pouring into Pakistan through the process of liberalization, in the effort to boost agricultural yields. It is also large landowners who have the surplus land available to offer to agri-businesses, as more effective land markets are created through ongoing donor-supported efforts like the computerization of land records. Thus, large landowners remain the main beneficiaries of the agricultural development processes as they are currently being propagated by donors like the World Bank, despite the pro-poor rhetoric of these donors. In their current form, programmatic interventions by donor agencies continue forwarding the agenda of opening up Pakistani agricultural markets to the mutual benefit of big businesses and the local elite. Prominent NGOs have also had little success in altering widespread socio-economic inequities by their involvement in tokenistic state land redistribution efforts or by implementing donor funded micro-credit programmes aiming to integrate poor households into the market economy.

Given this scenario, there are instances of resistance by poor farmers themselves, which have been supported by civil society, aiming to challenge the grip of powerful vested interests over land assets in rural Pakistan. The most recent resistance movement, launched by Pakistani farmers, involved landless tenants in the Punjab, who were threatened by eviction from military controlled farms. The movement itself was sparked due to military farms' attempts to maximize profits by altering the traditional land tenure system in favour of leasing contracts, motivated by the increasing trend of commercialization of land tenure arrangements. Despite trying to resist these trends, and challenging the military's control over agricultural land, the Anjuman Mazareen Punjab (AMP) movement has achieved limited success.

While the threat of eviction from military farms has temporarily been halted due to the resistance put up by the AMP, the movement has lost its cohesiveness, and its accomplishments are also not as significant as some of the literature on this movement suggests. There are evident tensions between the leadership of the AMP, especially the male and female representatives. Some of the emergent problems have to do with the male leadership trying to sideline female representatives within the movement, now that the goal of halting tenant eviction has been achieved. Other grievances are of a more personal nature, including the female leaders resenting the male leadership for excluding them from their due share of 'rewards' for having supported the AMP. The female representatives of the movement also made allegations of corruption among the male leadership concerning the use of AMP funds. Neither form of tensions within the AMP bodes well for the sense of cohesiveness that is of paramount important for such a grass-roots level movement.

The AMP remains embroiled in court cases, and its success against the military farm management remains partial. Given that contract farmers are also working on the military lands, and that the AMP has only managed to prevent the eviction of local tenant farmers rather than legally securing their ownership over the land they cultivate, indicates that they have not been able to effectively challenge the military's control over agricultural land in the country.

Hailing the AMP as a successful peasant movement capable of challenging the prevailing status quo is premature. However, there are useful lessons to be drawn from the inability of local and international NGOs in leveraging the solidarity created among tenants while resisting the might of the military, and turning the AMP into a sustainable social movement capable of challenging broader rural inequalities, especially the existing grip of the military over agricultural land. None of the NGOs which lent support to the AMP managed to effectively link the AMP to international movements aiming to represent the concerns of small farmers around the world.

Therefore, it is uncertain if the movement will be able to articulate a broader agenda for empowering poor farmers within the military farms and beyond. Even if the resistance by the AMP against the most powerful institution of the Pakistani state, the military, proves that poorer farmers do have some form of agency to act independently, the prevailing structures of power serve to limit the choices and opportunities available for poor farmers to achieve transformative change. This emerging experience of peasant resistance from

Pakistan needs closer attention by scholars and activists who place emphasis on indigenous peasant movements and NGO-supported transnational peasant movements, and their ability to reshape the prevailing global discourse concerning agricultural growth and rural poverty alleviation.

The elite-led agricultural development strategies of the Pakistani state thus have a significant amount of convergence with market-based donor prescriptions for achieving development, but their combined effects do little in terms of alleviating rural poverty in Pakistan. The market mechanism remains distorted in favour of larger landowners, and to the disadvantage of poor farmers, when it is applied in the particular political economy of rural Pakistan. The prevailing power nexus in Pakistan thus makes it very difficult to evade the elite capture of available development resources. Donors will not be able to effectively address such distortions either, unless they address a range of other power differentials due to which poor farmers continue to be marginalized. To correct existing structural inequalities, state and donor policy agendas need to be made more inclusive by strengthening the capacity of the excluded to participate on more equitable terms within the processes of production, distribution, and governance. The production process needs to be reformed to include poor farmers struggling to survive exclusively as wage earners or sharecroppers, by helping them become owners of productive assets like land. Effective land redistribution must also be accompanied by a range of institutional reforms, which place emphasis on pro-poor farming instead of capital and elite-led strategies for agricultural development.

Unless major revisions take place within donor and elite-dominated policy circles, their ability to address the glaring disparities in rural areas of countries like Pakistan remain slim. The hurdles preventing current resistance attempts from achieving meaningful change within the rural political economy does not mean that further unrest will not take place. The contours of this potential unrest, and its consequences, do, however, remain unpredictable.

Glossary

abiana irrigational water charge

acre a unit of land (equivalent to 0.405 hectares)

Baniya Hindu moneylender

batai share of crop given to landowner by sharecropper

dera a common meeting/resting place in rural areas, located on or near farmlands

Hari landless farmer/sharecropper

Haveli manor

Jagirdar land grant recipient

Jamadar labour force contractor/intermediary

Kamai member of the menial labourer caste

Kanal unit of land (equivalent 0.051 hectares)

Kharif autumn

Kissan farmer

Mansabdars land revenue officials under the Mughal Empire

Maurusi-hari permanent tenant

Mazdoor Labourer

Numberdar village official

Patwari land revenue official

Paulies weavers

Samindari land tenure system

Shariat Islamic law

Sufi Muslim mystic

Taccavi agricultural loan or advance

Tehsil administration unit situated within a district

Tehsildar Tehsil administrator

Toba well maker

Zamindar large landowners

References

Aftab, S. (2011). 'PTI economic plans: bold but unrealistic', *Friday Times*, 4 November, available from: www.thefridaytimes.com/beta2/tft/article.php?issue=20111104&page=4 (accessed 12 March 2012).

Agarwal, B. (1995). *A field of one's own: Gender and land rights in South Asia*, Cambridge: Cambridge University Press.

Ahmad, E. (1980). 'From potato sack to potato mash: The contemporary crisis of the third world', *Arab Studies Quarterly*, 2(3): 223–34.

Ahmad, E. (2007). 'The roots of violence', in Z. Mian and I. Ahmad (eds), *Making enemies, creating conflict: Pakistan's crises of state and society*, Lahore: Mashal Books.

Ahmed, A. and Gautam, M. (2013). *Increasing agricultural productivity*, Pakistan Policy Note 6, Washington, DC: World Bank.

Ahmed, F. (1984). 'Agrarian change and class formation in Sindh', *Economic and Political Weekly*, 19(39): 149–64.

Ahmed, F. (1996). 'Pakistan: ethnic fragmentation or national integration?', *Pakistan Development Review*, 35(4): 631–45.

Ahmed, M. (2011). 'Floods in Pakistan expose chronic poverty, and injustice', *Huffington Post*, available from: www.huffingtonpost.com/mubashir-ahmed/floods-in-pakistan-expose_b_843813.html (accessed 14 January 2013).

Ahmed, S. (2003). 'No land reforms anymore!', *Dawn*, 20 March, available from: http://archives.dawn.com/2003/03/20/op.htm (accessed 12 October 2010).

AHRC (2009). *Peasant courts must be established and the Sindh Tenancy Act, 1950 to be amended to end bonded labour*, available from: www.humanrights.asia/news/ahrc-news/AHRC-STM-040-2009 (accessed 15 November 2012).

Akhtar, A.S. (2006). 'The state as landlord in Pakistani Punjab: Peasant struggles on the Okara military farms', *Journal of Peasant Studies*, 33(3): 479–501.

Alavi, H. (1972). 'The state in post-colonial societies: Pakistan and Bangladesh', *New Left Review*, 1(74): 59–81.

Alavi, H. (1973). 'Elite farmer strategy and regional disparities in the agricultural development of Pakistan', *Economic and Political Weekly*, 8(13): 31–9.

Ali, I. (1987). 'Malign growth? Agricultural colonization and the roots of backwardness in the Punjab', *Past and Present*, 144: 110–32.

Ali, I. (1988). *The Punjab under imperialism, 1885–1947*, Princeton, NJ: Princeton University Press.

Ali, M. (2009). 'US aid to Pakistan and democracy: An overview', *Pakistan Journal of Social Sciences*, 29(2): 247–58.

Ali, Syed M. (2010a). *Landlessness in Pakistan: A policy brief*, Lahore: Pakistan Policy Group.

Ali, Syed M. (2010b). 'Failure to address skewed landownership in the developing world', *Traffic*, 12: 57–72.

Ali, Syed M. (2014). Bypassing poor farmers: Market based approaches to agriculture in rural Pakistan', *Melbourne Journal of Politics*, 39: 19–39.

Ali, T. (2006). 'Bought with western cash', *Guardian*, 7 April, available from: www.guardian.co.uk/commentisfree/2006/apr/07/pakistan (accessed 19 February 2012).

Altaf, Z. (2010). 'Food security in pluralistic Pakistan', in M. Kugelman and R. Hathaway (eds), *Hunger pains: Pakistan's food insecurity*, Washington, DC: Woodrow Wilson Center.

Anwar, T., Qureshi, S.K., and Ali, H. (2004). 'Landlessness and rural poverty in Pakistan', *Pakistan Development Review*, 43(3): 885–904.

Arif, M. (2008). *Land rights: Peasants' economic justice*, Lahore: South Asia Partnership-Pakistan.

Asian Development Bank & World Bank (2010). *Pakistan floods 2010: Preliminary damage and needs assessment*, Islamabad: World Bank & Asian Development Bank.

Associated Press of Pakistan (2012). 'Over 60 per cent of Pakistani lawmakers evade taxes', *Dawn*, 12 December, available from: http://dawn.com/2012/12/12/over-60-per-cent-of-pakistani-lawmakers-evade-taxes-report/ (accessed October 2012).

Bano, M. (2012). *Breakdown in Pakistan: How aid is eroding institutions for collective action*, Stanford, CA: Stanford University Press.

Barbier, E. (2000). 'Links between economic liberalization and rural resource degradation in the developing regions', *Agricultural Economics*, 23(2): 299–310.

Bateman, M. (2010). *Why microfinance doesn't work*, London: Zed Books.

Benjamin, S., Bhuvaneswari, R., and Rajan, P. (2007). *Bhoomi: 'E-Governance', or, an anti-politics machine necessary to globalize Bangalore?* Bangalore: Collaborative for the Advancement of Studies in Urbanism through Mixed Media.

Besley, T. and Burgess, R. (2000). 'Land reform, poverty reduction, and growth: Evidence from India', *Quarterly Journal of Economics*, 115(2): 389–420.

Birner, R. and Resnick, D. (2010). 'The political economy of policies for smallholder agriculture', *World Development*, 38(10): 1442–52.

Board of Investment. (2004). *Salient features of investment policy for corporate agriculture farming*, Islamabad: Prime Minister's Secretariat.

Borras, S. (2006). 'The Philippine land reform in comparative perspective: Some conceptual and methodological implications', *Journal of Agrarian Change*, 6(1): 69–101.

Borras, S. (2008). *Competing views and strategies on agrarian reform. Volume I: International Perspective*, Manila: Ateneo De Manila University Press.

Borras, S. (2009). 'Agrarian change and peasant studies: changes, continuities and challenges – an introduction', *Journal of Peasant Studies*, 36(1): 5–31.

Borras, S., Edelman, M., and Kay, C. (2008). 'Transnational agrarian movements: Origins and politics, campaigns and impact', *Journal of Agrarian Change*, 8(2): 169–204.

Borras, S.M. (2009). 'Agrarian change and peasant studies: changes, continuities and challenges – an introduction', *Journal of Peasant Studies*, 36(1): 5–31.

Borras, S.M. and Franco, J.C. (2010). 'Contemporary discourses and contestations around pro-poor land policies and land governance', *Journal of Agrarian Change*, 10(1): 1–32.

Brohi, N. (2010). *Gender and land reforms in Pakistan*, Islamabad: Sustainable Development Policy Institute.

Brooks, D., Hasan, R., Lee, J., and Son, H. (2010). 'Closing development gaps: Challenges and policy options', *Asian Development Review*, 27(2): 1–28.

Byres, T. and Bernstein, H. (2001). 'From peasant studies to agrarian change', *Journal of Agrarian Change*, 1(1): 1–57.

Cecelia, B. (2008). *Subprime lending: Lessons for the microfinance industry*, Concord: Microcapital.

Cellini, F., Chesson, A., Colquhoun, I., Constable, A., Davies, H.V., Engel, K.H., Gatehouse, A.M.R., Kärenlampi, S., Kok, E.J., Leguay, J.J., Lehesranta, S., Noteborn, H.P.J.M., Pedersen, J., and Smith, M. (2004). 'Unintended effects and their detection in genetically modified crops', *Food and Chemical Toxicology*, 42(7): 1089–1125.

Centre for Good Governance (2002). *Computer-aided administration of registration department (CARD) review*, Andhra Pradesh: Centre for Good Governance.

Chaudhry, I., Malik, S., and Ashraf, M. (2006). 'Rural poverty in Pakistan: Some related concepts, issues and empirical analysis', *Pakistan Economic and Social Review*, 44(2): 259–76.

Chaudhry, J.R.H.a.M.G. (1974). 'The 1972 land reforms in Pakistan and their economic implications: A preliminary analysis', *Pakistan Development Review*, 13.

Chaudry, G. (1999). 'The theory and practice of agricultural income tax in Pakistan and a viable solution', *Pakistan Development Review*, 38(4): 757–68.

Chawla, R. (2004). 'Online delivery of land titles to rural farmers in Karnataka, India', *Scaling Up Poverty Reduction: A Global Learning Process and Conference*, Shanghai: World Bank.

Cheema, A., Khalid, L., and Patnam, M. (2008). 'The geography of poverty: Evidence from the Punjab', *Lahore Journal of Economics*, special edition, 163–88.

Cheema, A., Mohmand, S., and Patnam, M. (2009). *Colonial proprietary elites and institutions: Persistence of de facto political control*, Brighton: Institute of Development Studies.

Cheema, U. (2012). *Representation without taxation: An analysis of MPs income tax returns 2011*, Islamabad: Centre for Peace and Development Initiatives.

Cheesman, D. (1996). *Landed power and rural indebtedness in colonial Sind*, Surrey: Curzon Press.

CIPR (2002). 'Agriculture and genetic resources', in CIPR (ed.), *Integrating intellectual property rights and development policy*, London: CIPR.

Davidson, A., Ahmad, M., and Ali, T. (2001). *Dilemmas of agricultural extension in Pakistan: Food for thought*, London: Agricultural Research and Extension Network.

Dawn (2010). 'Court frees 78 bonded labourers', *Dawn*, 4 March, available from: archives.dawn.com/archives/136286 (accessed April 2011).

DFID (2009). *Punjab economic opportunities programme (PEOP): Programme document*, Islamabad: DFID-Pakistan.

Dorosh, P. and Salam, A. (2007). *Distortions to agricultural incentives in Pakistan*, Washington, DC: World Bank.

Easterly, W. (2001). *The elusive quest for growth: Economists' adventures and misadventures in the tropics*, Cambridge, MA: MIT Press.

Eglar, Z. (2010). *A Punjabi village in Pakistan: Perspectives on community, land and economy*, Karachi: Oxford University Press.

FBR (2012). *A review of resource mobilization efforts of the Federal Board of Revenue*, Islamabad: FBR.

Ferguson, J. and Lohmann, L. (1994). 'The anti-politics machine development and bureaucratic power in Lesotho', *The Ecologist*, 5(24): 176–81.

FIRMS (2010). *FIRMS project: Annual progress report*, Lahore: United States Agency for International Development (USAID) FIRMS Project.

Forster, R. and Schnell, S. (2003). *Participation in monitoring and evaluation of PRSPs: A document review of trends and approaches emerging from 21 Full PRSPs*, Washington, DC: World Bank.

Gardezi, H. and Rashid, J. (1983). *Pakistan the unstable State*, Lahore: Vanguard Books.

Gazdar, H. (2005). *Determinants and drivers of poverty reduction in rural Pakistan*, Islamabad: Asian Development Bank.

Gazdar, H. (2007). *Rural economy and livelihoods in Pakistan, determinants and drivers of poverty reduction and ADB's contribution in rural Pakistan*, Islamabad: Asian Development Bank.

Gazdar, H. (2009). *Policy responses to economic inequality in Pakistan*, Islamabad: United Nations System in Pakistan.

Gazdar, H. (2011). 'The fourth round, and why they fight on: The history of land reform in Pakistan', in R. Kalshian (ed.), *Leveling the playing field: A survey of Pakistan's land reforms*, Nepal: Panos South Asia.

Gera, N. (2004). 'Food security under structural adjustment in Pakistan: Asian Survey', *Asian Survey*, 44(3): 358–68.

Ghaus-Pasha, A. and Iqbal, M.A. (2003). *Defining the nonprofit sector: Pakistan*, Baltimore, MD: Johns Hopkins Center for Civil Society Studies.

Giddens, A. (1991). *Sociology*, Oxford: Polity Press.

Gilens, M. (1999). *Why Americans hate welfare: Race, media, and the politics of anti-poverty policy*, Chicago, IL: Chicago Press.

Gillespie, A. and Michelson, M.R. (2011). 'Participant observation and the political scientist: Possibilities, priorities, and practicalities', *Political Science and Politics*, 44(2): 261–5.

Given, L. (2008). *The Sage encyclopaedia of qualitative research methods*. Thousand Oaks, CA: Sage.

Government of Pakistan (1973). *Constitution of Pakistan*. Islamabad: Government of Pakistan.

Government of Pakistan (2003). *Accelerating economic growth and reducing poverty: The road ahead* [Pakistan' poverty reduction strategy paper], Islamabad: Ministry of Finance.

Government of Pakistan (2005). *Mid-term development framework (MTDF) 2005–2015*. Islamabad: Planning Commission.

Government of Pakistan (2008). *Poverty reduction strategy paper (PRSP)-II*. Islamabad: Ministry of Finance.

Government of Pakistan (2009a). *Pakistan economic survey 2008–09*. Islamabad: Ministry of Finance.

Government of Pakistan (2009b). *Pakistan employment trends for women*, Islamabad: Ministry of Labour and Manpower.

Government of Pakistan (2010). *Labour policy of Pakistan 2010*, Islamabad: Ministry of Labour and Manpower.

Government of Punjab (1958). *Punjab agriculturalists loan act of 1958*, Lahore: Punjab Assembly.

Government of Punjab (2013). *The Punjab gazette minimum wages 2013*, Notification no. SO(DII)MW/2011(P-II), Lahore: Government of Punjab.

Government of Sindh (2008a). *Grant of state land to landless haris*, Karachi: Government of Sindh.

Government of Sindh (2008b). *Ushering structural transformation in Sindh; Government of Sindh's program for grant of state land to poor landless haris*, Karachi: Government of Sindh.

Government of Sindh (2012). *Population of Sindh*, Karachi: Population welfare department website, available from: http://www.pwdsindh.gov.pk (accessed 15 January 2014).

Government of Sindh (2013). *Minimum wages for unskilled workers 2013*, Notification no. L-II/13–14/78-I, Karachi: Government of Sindh.

Gupta, A. (1995). 'Blurred boundaries: The discourse of corruption, the culture of politics, and the imagined state', *American Ethnologist*, 22(2): 375–402.

Habib, I. (1983). 'The peasant in Indian history', *Social Scientist*, 11(3): 21–64.

Habib, I. (1999). *The agrarian system of Mughal India 1526–1707*, New Delhi; New York: Oxford University Press.

Haider, S. (2013). Rigging: 49 per cent polling stations across Pakistan received around 100 pc turnout, *Dawn*, 14 May, available at: www.dawn.com/news/1027132/rigging-49-polling-stations-received-over-100pc-turnout-across-pakistan.

Haq, A. and Khalid, Z. (2011). *Assessing risks to microfinance in Pakistan: Findings from a risk assessment survey*, Islamabad: Pakistan Microfinance Network.

Harrell, M. and Bradley, M. (2009). *Data collection methods: Semi-structured interviews and focus groups*, Santa Monica, CA: RAND Corporation.

Harvey, H. (2003). *The new imperialism* Oxford: Oxford University Press.

Hasan, P. (1997). 'Learning from the past: A fifty-year perspective on Pakistan's development', *Pakistan Development Review*, 36(4): 355–402.

Hasnain, K. (2010). 'CM's land lease scheme: BoR orders inquiry into allotments in Bhakkar', *Dawn*, available from: http://dawn.com/2010/09/30/cm-s-land-lease-scheme-bor-orders-inquiry-into-allotments-in-bhakkar/ (accessed 19 February 2012).

Hassan, M.U. (2008). 'The conception, design and implementation of IMT in Pakistan's Punjab: A public policy reflection', *International Conference on Water Resource Policy in South Asia*, Colombo, Sri Lanka South Asia Consortium for Interdisciplinary Water Resources Studies (SaciWATER).

Herring, R. (1979). 'Zulfikur Ali Bhutto and the "eradication of feudalism" in Pakistan', *Comparative Studies in Society and History*, 21(4): 519–57.

Herring, R. J and Kennedy, C.J. (1979). 'The political economy of farm mechanization policy: Tractors in Pakistan', in E.A. Raymond Hopkins (ed.), *Food): olitics and agricultural development*, Boulder, CO: Westview.

Herring, R.J. (1983). *Land to the tiller: The political economy of agrarian reform in South Asia*, New Haven, CT: Yale University Press.

Hisam, Z. (2007). *Denial and discrimination: Labour rights in Pakistan*, Karachi: Pakistan Institute of Labour Education and Research.

Hulme, D. (2003). 'Microfinance: Evolution, achievements, and challenges', in M. Harper (ed.), *Microfinance: evolution, achievements, and challenges*, London: ITDG.

Hussain, A. (1984). 'Land reforms in Pakistan: A reconsideration', *Bulletin of Concerned Asian Scholars*, January–March, Colorado.

Hussain, A. (2005). 'Rising poverty trends: Causes and remedies', in IFPR (ed.), *The role of agriculture in poverty reduction in Pakistan*, Lahore, Pakistan: International Food Policy Research Institute and Beacon House National University.

Hussain, A. (2008). 'Institutional imperatives of poverty reduction', *Institute of Public Policy*, Lahore: Beaconhouse National University.

Hussain, A., Kemal, A., Hamid, A., Ali, I., Mumtaz, K., and Qutub, A. (2003). *Poverty, growth and governance*, Karachi: Oxford University Press.

Hussein, M. (2009). *State of microfinance in Pakistan*, Dhaka: Institute of Microfinance.

Hussein, M., Saleemi, A., Malik, S. and Hussain, S. (2003). *Bonded labor in agriculture: A rapid assessment in Sindh and Balochistan, pakistan*, Geneva: International Labour Organization.

Hussein, M., Saleemi, A., Malik, S., and Hussain, S. (2004). *Bonded labor in agriculture: A rapid assessment in Sindh and Balochistan, Pakistan*, Geneva: International Labour Organization.

IFAD (2013). *Down to earth: Sustainable rural transformation*, Rome: IFAD.

ILO (1973). *Minimum age convention (C 138)*, Geneva: ILO.

IPO-Pakistan (2012). *Plant Breeders Rights Bill 2012*, Islamabad: International Property Organization (IPO) of Pakistan, Government of Pakistan.

Jacoby, H., Murgai, R., and Rehman, S.U. (2004). 'Monopoly power and distribution in fragmented markets: The case of groundwater', *Review of Economic Studies*, 71(3): 783–808.

Jalal, A. (1995). *Democracy and authoritarianism in South Asia: A comparative and historical perspective*, New York: Cambridge University Press.

Kamal, S. (2009). *Use of water for agriculture in Pakistan: Experiences and challenges*, Lincoln, NE: University of Nebraska.

Karim, M. (2014). 'Fight on for land', *The News*, 25 May, available from: http://tns.thenews.com.pk/category/dialogue/ (accessed 15 June).

Kaufmann, D., Kraay, D., and Mastruzzi, M. (2009). *Governance matters 2009: Learning from over a decade of the worldwide governance indicators*, Washington, DC: Brookings Institution.

Kerr, S. and Bokhari, F. (2008). 'UAE investors buy Pakistan farmland', *Financial Times*, 11 May 2008, available from: www.ft.com/intl/cms/s/0/c6536028-1f9b-11dd-9216-000077b07658.html#axzz362u5jIuY (accessed 15 January 2014).

Khan, S.R. (1999). *Do World Bank and IMF policies work?* London: Macmillan.

Khan, S.R. and Yusuf, M. (2004). *Potential impact on southern farmers of reducing northern subsides: Reflections from Pakistan*, Islamabad: Sustainable Development Policy Institute.

Khan, Z. (2011). 'National Assembly passes landmark women's rights bill', *Express Tribune*, 15 November, available from: http://tribune.com.pk/story/292165/prevention-of-anti-women-practices-bill-unanimously-approved-by-na/ (accessed 20 March 2012).

Kosambi, D.D. (1975). *An Introduction to the study of Indian history*, Bombay: Popular Prakashan.

Kugelman, M. and Hathaway, R. (eds) (2010). *Hunger pains: Pakistan's food insecurity*, Washington, DC: Woodrow Wilson International Center for Scholars.

Kulavuz-Onal, D. (2011). 'Voicing the less heard: A review of focus group methodology: Principles and practice', *Qualitative Report*, 16: 1743–8.

Kvale, S. (1996). *An introduction to qualitative research interviewing*, Thousand Oaks, CA: Sage.

La Via Campesina. (2004). *Violations of peasants' human rights: A report on cases and patterns of violation*, Honduras: La Via Campesina.

Lancaster, J. (2003). 'Pakistan's modern feudal lords', *Washington Post*, 8 April.

Lieten, K. and Breman, J. (2002). 'A pro-poor development project in rural Pakistan; an academic analysis and a non-intervention', *Journal of Agrarian Change*, 2(3): 331–55.

Lindemann, C. (2010). '"Landless peasant" activism in Brazil: Fighting for social inclusion through land reform', Ph.D. thesis, University of Melbourne.

Lodhi, H. and Cristóbal, K. (2010). 'Surveying the agrarian question (part 1): unearthing foundations, exploring diversity', *Journal of Peasant Studies*, 37(1): 177–203.

Lorenzo, C., Vermeulen, S., Leonard, R., and Keeley, J. (2009). *Land grab or development opportunity? Agricultural investment and international land deals in Africa*, Rome: Food and Agriculture Organization.

Mahbub-ul-Haq Human Development Centre (MHHDC) (2003). *Human development in South Asia: The gender question*, Karachi: Oxford University Press.

Mahmood, A., Sheikh, A.D., and Akmal, N. (2010). 'Impact of trade liberalization on agriculture in Pakistan', *Journal of Agricultural Research*, 28(1): 121–30.

Malik, N. (2008). 'Reply: What's changed (since 1975)?', *Dialect Anthropology*, 32: 31–7.

Malik, N. (2009). 'The modern face of traditional agrarian rule: local government in Pakistan', *Development in Practice*, 19(8): 997–1108.

Mansuri, G. and Jacoby, H. (2006). *Incomplete contracts and investment: A study of land tenancy in Pakistan*, Washington, DC: World Bank.

Martin-Prével, A. (2014a). *Corporatising agriculture: World Bank's rankings facilitate land grabs*, London: Brettonwoods Project.

Martin-Prével, A. (2014b). *Willful blindness: How World Bank's country rankings impoverish smallholder farmers*, Oakland, CA: Oakland Institute.

Marx, K. (1963). *The eighteenth brumaire of Louis Bonaparte*, New York: International Publishers.

McMichael, P. (2009). 'Banking on agriculture: A review of the *World Development Report 2008*', *Journal of Agrarian Change*, 9: 235–46.

Mercer, C. (2002). 'NGOs, civil society and democratization: a critical review of the literature', *Progress in Development Studies*, 2(1): 5–22.

MHHDC (2009). *Human development in South Asia 2009: Trade and human development in South Asia*, Karachi: Oxford University Press.

Minto, A. and Minto, B. (2012a). *Electoral reforms petition under article 184(3) of the Constitution of Pakistan 1973*, Islamabad.

Minto, A. and Minto, B. (2012b). *Land reforms petition under article 184(3) of the Constitution of Pakistan 1973*, Islamabad.

Mitrany, D. (1952). *Marx against the peasant: a study in social dogmatism*, 2nd edn, London: Weidenfeld and Nicolson.

Mitray, S., Mookherjeez, D., Torerox, M., and Visaria, S. (2012). *Asymmetric information and middleman margins: An experiment with West Bengal potato farmers*, Washington, DC.

MQM (2010). *Redistributive land reforms bill 2010*, Karachi: MQM.

Mumtaz, K. and Mumtaz, S. (2012). 'Women's participation in the Punjab peasant movement: From community rights to women's rights? Political conflict and women in South Asia', *South Asia Journal*, 35: 138–50.

Murray, U. and Hurst, P. (2009). *Mainstreaming responses for improvement of the girl child in agriculture*, Geneva: International Labour Organization.

National Coalition against Bonded Labour (2009). *The state of bonded labour in Pakistan*, Islamabad.

NDMA (2012). *Disaster risk management needs report*, Islamabad: National Disaster Management Agency.

NDMA (2013). *Summary of damages in 2012 Floods*, Islamabad: National Disaster Management Agency.

Niazi, T. (2004). 'Rural poverty and the green revolution: The lessons from Pakistan', *Journal of Peasant Studies*, 31(2): 242–60.

NIPS (2008). *Demographic and health survey*, Islamabad: National Institute of Population Studies.

Oxfam International (2004). *From 'donorship' to ownership? Moving toward PSRP round two*, briefing Paper 54, London: Oxfam International.

Pakistan Bureau of Statistics (2010). *Agricultural census 2010 – Pakistan report*, Islamabad: Government of Pakistan.

Pakistan Press International (2010). 'MQM's land reforms bill un-Islamic: JUP', *Express Tribune*, 13 October, available from: http://tribune.com.pk/story/62096/mqms-land-reforms-bill-un-islamic-jup/ (accessed 11 May 2012).

PDI. (2009). *Sindh Government's land distribution program: Issues and challenges*, Karachi: Participatory Development Initiatives.

Pereira, J.M.M. (2005). *From panacea to crisis: Grounds, objectives and results of the World Bank's market-assisted land reform in South Africa, Colombia, Guatemala and Brazil*, Rio de Janeiro: Land Action.

PIDE (2012). 'Rural poverty dynamics in Pakistan', *PIDE Viewpoint*, October issue, Islamabad: Pakistan Institute of Development Economics.

PML-N (2013), *National Agenda for Real Change, Manifesto 2013*, available from: www.pmln.org/pmln-manifesto-english/ (accessed May 2013).

PPP (1970). *Manifesto*, available from: www.ppp.org.pk/manifestos/1970.html (accessed 12 November 2012).

PPP (2008). *Towards peace and prosperity in Pakistan – Manifesto 2008*, available from: www.ppp.org.pk/manifestos/2008.pdf (accessed 12 November 2012).

Punjab Bureau of Statistics (2011). *Statistical pocket-book of the Punjab*, Lahore: Government of Punjab.

Qazi, U. (2005). *Computerization of land records in Pakistan*, Islamabad: Leadership in Environment and Development – Pakistan (LEAD-Pakistan).

Qureshi, M. and Qureshi, S.K. (2004). 'Impact of changing profile of rural land market in Pakistan on resource allocation and equity', *Pakistan Development Review*, 43(4): 471–92.

Rabionet, S.E. (2011). 'How I learned to design and conduct semi–structured interviews: An ongoing and continuous journey', *Qualitative Report*, 16(2): 563–6.

Rajayyan, K. (1971). *South Indian rebellion: The first war of independence, 1800–1801*, Mysore: Rao and Raghavan.

Rana, A. (2011). 'Scandal at NFML: Ministers allegedly involved in Rs300b fertiliser fraud', *Express Tribune*, 29 December, available from: http://tribune.com.pk/story/313450/scandal-at-nfml-ministers-allegedly-involved-in-rs300b-fertiliser-fraud/ (accessed 11 March 2012).

RDI (2009). *Women's inheritance rights to land and property in South Asia*, Washington, DC: RDI.

Reis, E. and Moore, M. (2005). 'Elite, perceptions and poverties', in E. Reis and M. Moore (eds), *Elite perceptions of poverty and inequality*, London: Zed Books.

Revenue Department (2012). *Land records management and information system (LRMIS)*, Government of Punjab, available from: www.punjab-zameen.gov.pk/details. php?menuid=2&submenuid=8 (accessed 18 November 2012).

Rossett, P. (2001). *Tides shift on agrarian reform*, Oakland, CA: Institute of Food and Development Policy.

RSPN (2012). *RSPN donors*: RSPN website, available from: www.rspn.org/donors/ donors.html (accessed 17 January 2013).

Sadeque, N. (2009). 'Giving away the family silver', *Newsline Magazine*, 26 October.

Sadeque, N. (2012). 'Time for meaningful land reform', *Express Tribune*, 21 May, available from: http://tribune.com.pk/story/382143/time-for-meaningful-land-reform/ (accessed 11 February 2013).

Salim, A. (2008). *Peasant land rights movements of Pakistan*, Islamabad: Sustainable Development Policy Institute.

Schuler, P. (2004). *Pakistan tariff rationalization study*, Islamabad: World Bank.

SDPI (2008). *Land rights for Muslim women: Review of law and policy*, Islamabad: SDPI.

SDPI (2008). *Land rights for Muslim women: Review of law and policy*, Islamabad: SDPI.

Séralini, G. E., Cellier, D., and Vendomois, J. (2007). 'New analysis of a rat feeding study with a genetically modified maize reveals signs of hepatorenal toxicity', *Archives of Environmental Contamination and Toxicology*, 52(4): 596–602.

Settle, A. (2013), *Agricultural land acquisition by foreign investors in Pakistan*, London: Land Deal Politics Initiative and the Institute for Development Studies.

Shafique, D.M.S. (2008). 'Potential options for effective agricultural extension in Pakistan', *International Hydro Politics*, 19 November, available from: www.international hydropolitics.com/node/2.

Siddiqa, A. (2006). 'The new land barons?', *Newsline*, Karachi: Newsline Publications.

Siddiqa, A. (2007). *Military Inc.: Inside Pakistan's military economy*, London: Pluto.

Singh, S. (2006). *Corporate farming in India: Is it must for agricultural development?* Ahmadabad: Indian Institute of Management.

Skogly, S. (2001). *The human rights obligations of the World Bank and the International Monetary Fund*, London: Cavendish.

SPARC (2007). *State of Pakistan's children – 2006*, Islamabad: SPARC.

Stein, H. (2010). 'World Bank agricultural policies, poverty and income inequality in sub-Saharan Africa', *Cambridge Journal of Regions, Economy and Society*, 4(1): 79–90.

Steinbeck, J. (1939). *The grapes of wrath*, New York: Viking.

Stiglitz, J. (2002). 'Participation and development: Perspectives from the comprehensive development paradigm', *Review of Development Economics*, 6(2): 163–82.

Stokes, E. (1978). *The peasant and the Raj: Studies in agrarian society and peasant rebellion in colonial India*, Cambridge: Cambridge University Press.

Stokes, E. (1989). *The English utilitarians in India*, Delhi: Oxford University Press.

Supreme Court of Pakistan (2012). *Judgment on Constitution Petition No.87 of 2011; Constitution Petition challenging election campaigns expenses regulation case*, Islamabad.

Syngenta (2012). *Key facts about Syngenta*, available from: www.syngenta.com/global/ corporate/en/news-center/company-profile/Pages/key-facts.aspx (accessed 11 August 2012).

Talbot, I. (2007). 'Punjab under colonialism: Order and transformation in British India', *Journal of Punjab Studies*, 14(1): 3–10.

Tavernise, S. (2010). 'Upstarts chip away at power of Pakistani elite', *New York Times*, 28 August, available from: www.nytimes.com/2010/08/29/world/asia/29feudal.html? pagewanted=all&_r=0 (accessed 19 February 2013).

The News (2011). 'Female farmers deprived of their right', 9 March, available from: www.thenews.com.pk/Todays-News-6–35230-Female-farmers-deprived-of-their-rights (accessed 5 April 2012).

Thorner, D. *et al.* (eds) (1986). *Chayanov: The theory of peasant economy*, Madison, WI: University of Wisconsin Press.

Tirmizi, F. (2012). 'High connections: David Miliband joins Pakistani private equity firm', *Express Tribune*, 21 January, available from: http://tribune.com.pk/story/324941/high-connections-david-miliband-joins-pakistani-private-equity-firm (accessed 19 April 2012).

Toor, S. (2010). 'The structural dimensions of food insecurity in Pakistan', in M. Kugelman and R. Hathaway (eds), *Hunger pains: Pakistan's food insecurity*, Washington, DC: Woodrow Wilson Center for Scholars.

Toor, S. (2011). *The state of Islam: Culture and cold war politics in Pakistan*, New York: Palgrave Macmillan.

UN (1992). *Convention on biological diversity*, New York: United Nations.

UNOCHA (2011). *Pakistan floods: Rapid response plan*, Islamabad: UNOCHA.

USAID (2010). *Land tenure, property rights and insurgency in Pakistan*, Islamabad: USAID.

Walsh, D. (2010). 'Pakistan's flood-ridden lands are crying out for political change – but can Jamshed Dasti bring it?' *Guardian*, 3 October, available from: www.guardian.co.uk/world/2010/oct/03/pakistan-flood-jamshed-dasti-politics (accessed 16 October 2012).

World Bank (2000). *Structural adjustment, economic growth and the aid-debt service system: The least developed countries report*, Washington, DC: World Bank.

World Bank (2002). *Sindh structural adjustment credit project*, Islamabad: World Bank.

World Bank (2005). *Bridging the gender gap: opportunities and challenges*, Washington, DC: World Bank.

World Bank (2006). *Land Records Management and Information Systems Program (LRMIS-P) Province of Punjab*, Washington, DC: World Bank.

World Bank (2007a). *Pakistan Promoting Rural Growth and Poverty Reduction*, Washington, DC: World Bank.

World Bank (2007b). *World development report 2008: Agriculture for development*, Washington, DC: World Bank.

World Bank (2008). *Leveling the playing field in international agricultural trade*, Washington DC: World Bank.

World Bank (2013). *Benchmarking the Business of Agriculture: Project Overview*, World Bank website, available from: http://bba.worldbank.org/project-overview (accessed 2014).

World Bank (2014). *Pakistan: Country snapshot*, Islamabad: World Bank.

WTO (1995). *Trade-Related Aspects of Intellectual Property Rights (TRIPs) Agreement*, Geneva: World Trade Organization.

Zaidi, A. (2001). *The economy, poverty and bonded labour in Pakistan*, Karachi: Pakistan Institute of Labour Education and Research.

Zaidi, R.Z. (2012). 'Entering the field: Examining the relevance of political ecology to the agrarian struggle of Anjuman Muzareen Punjab in Pakistan', *Singapore Journal of Tropical Geography*, 33(1): 63–76.

Zaidi, S.A. (2012a). 'The captivating vision of the "New growth strategy": The missing political economy perspective', *Lahore Journal of Economics*, 17: 33–49.

Zaidi, S.A. (2012b). 'Entering the field: Examining the relevance of political ecology to the agrarian struggle of Anjuman Muzareen Punjab in Pakistan', *Singapore Journal of Tropical Geography*, 33(1): 63–76.

Zhuang, J., Dios, E., and Lagman-Martin, A. (2010). 'Governance and institutional quality and the links with growth and inequality: How Asia fares', in J. Zhuang (ed.), *Poverty, inequality, and inclusive growth in Asia: Measurement, policy issues, and country studies*, London: Anthem Press for the Asian Development Bank.

Index